GUARDED HOT PLATE AND HEAT FLOW METER METHODOLOGY

A symposium
sponsored by
ASTM Committee C-16 on
Thermal Insulation and
The National Research
Council of Canada
Quebec, Can., 7, 8 Oct. 1982

ASTM SPECIAL TECHNICAL PUBLICATION 879
C. J. Shirtliffe, National Research Council
of Canada, and
R. P. Tye, Dynatech R/D Company, editors

ASTM Publication Code Number (PCN)
04-879000-10

1916 Race Street, Philadelphia, PA 19103

Library of Congress Cataloging-in-Publication Data

Guarded hot plate and heat flow meter methodology.

(ASTM special technical publication; 879)
"ASTM publication code number (PCN) 04-879000-10."
Includes bibliographies and index.
1. Insulation (Heat)—Thermal properties—
Congresses. 2. Heat—Transmission—Measurement—
Equipment and supplies—Congresses. I. Shirtliffe,
C. J. II. Tye, R. P. (Ronald Phillip) III. ASTM
Committee C-16 on Thermal Insulation. IV. National
Research Council of Canada. V. Series.
TH1715.G8 1985 621.402′4 85-18531
ISBN 0-8031-0423-5

Library of Congress Catalog Card Number: 85-18531

NOTE

The Society is not responsible, as a body,
for the statements and opinions
advanced in this publication.

Printed in Ann Arbor, MI
October 1985

Foreword

The papers in this publication, *Guarded Hot Plate and Heat Flow Meter Methodology*, were presented at a symposium held in Quebec, Canada, 8, 9 October 1982. The symposium was sponsored by ASTM Committee C-16 on Thermal Insulation and the National Research Council of Canada. C. J. Shirtliffe, National Research Council of Canada, and R. P. Tye, Dynatech R/D Company, are editors of this publication.

Related ASTM Publications

Thermal Insulation, Materials, and Systems for Energy Conservation in the '80s, STP 789 (1983), 04-789000-10

Thermal Insulation Performance, STP 718 (1980), 04-718000-10

Thermal Transmission Measurements of Insulation, STP 660 (1979), 04-660000-10

Thermal Insulations in the Petrochemical Industry, STP 581 (1975), 04-581000-10

Heat Transmission Measurements in Thermal Insulations, STP 544 (1974), 04-544000-10

A Note of Appreciation to Reviewers

The quality of the papers that appear in this publication reflects not only the obvious efforts of the authors but also the unheralded, though essential, work of the reviewers. On behalf of ASTM we acknowledge with appreciation their dedication to high professional standards and their sacrifice of time and effort.

ASTM Committee on Publications

ASTM Editorial Staff

Contents

Introduction

ASTM Committee C16 on Thermal Insulations will be celebrating its golden anniversary in 1988. During the early days of its establishment over forty years ago the membership realized the significance and importance of the need to develop a reliable test method to measure the thermal performance of the materials and systems for which they were responsible. Much attention was paid to this subject, and, as a result of their early efforts, ASTM C 177 describing and utilizing the guarded hot plate method was published in 1945. This quickly became recognized and accepted as the fundamental absolute technique for measuring the thermal conductivity of thermal insulations and materials of low thermal conductivity. Many countries subsequently produced similar types of test method documents based on the technique.

During the following two decades there were significant developments of newer types of thermal insulations including fibrous glass and cellular plastics. In addition, new applications and uses of thermal insulation combined with increased use, particularly in buildings, produced a significant growth in the volume of insulation being manufactured and used. There was corresponding increase in the need for the measurements of properties of these materials and especially the thermal performance.

It became obvious however, that there was now a requirement for a method of measurement of thermal conductivity which was both more rapid and less complex than that utilizing the guarded hot plate, particularly for use during the manufacturing process. During this era several organizations had started working with a heat flux transducer calibrated within a hot and cold plate system with a reference material of known thermal properties. Their experience resulted first in a tentative test method in 1962 and ultimately in 1964 with ASTM C 518 the Heat Flow Meter Method. This again became accepted worldwide as the prime secondary method for measuring thermal conductivity of thermal insulations.

The first documents were, in general, directed towards evaluation of such materials at or near room temperature. Their contents were based totally on the expertise and experience of those few careful and dedicated experimental workers who had the foresight to become members of the T-6 Subcommittee of C16. This was the forerunner of the present C16.30 Subcommittee on Thermal Measurements, the body which organized the present meeting. However, as time progressed, the requirements for measurements at both elevated and cryogenic temperatures became more necessary. Thus, in 1963 the first major revision of C 177 was undertaken to reflect this need with a further more detailed revision taking place in 1971. In 1970, the C 518 method was also revised to reflect the additional knowledge gained in using the technique since it was first adopted. As a result of these revisions it was believed then that both methods reflected the state of the art and provided the means to obtain reliable results on the materials for which they were suitable.

However, as we are all aware, the world energy crisis of the early seventies provided a significant stimulus to the concept of energy conservations both for building and industrial applications. As a result, more insulation was prescribed, thicknesses of use increased, many more people became involved in measurement both in the laboratory and production areas, more and larger equipment, including commercial models, were being developed and used, and regulatory bodies became involved to mandate and regulate performance requirements. Each of these factors imposed new burdens on the measurement community.

It quickly became apparent that heat transmission characteristics particularly radiation and to a lesser extent convection imposed a need for more stringent test criteria. Heat transfer phenomena were complex through a media where solid conduction was not the prime mode of heat transmission. The old simple definition of thermal insulation having a thermal conductivity was abandoned in favor of the more correct individually measured thermal resistance or thermal conductance concept. In 1976, both C 177 and C 518 documents were revised radically in their philosophy to reflect the true requirements of the thermal insulation community.

During the seventies, the C16.30 Thermal Measurements Subcommittee took the world lead in this area by organizing three symposia. These were directed towards general and specific understanding of heat transmission phenomena in insulations and their measurement. Position Papers were developed by the Subcommittee to address the specific areas of measurement philosophy and the need for reference materials and transfer standards for ensuring good measurements. International participation was sought and encouraged. This stimulated significant cooperative efforts between North America and European workers, and overseas membership of the subcommittee grew. Members of the Subcommittee actively participated in the formation in 1976 of the ISO TC163 Committee on Thermal Insulations and in its subsequent work, particularly on the Subcommittee SCI Test Method Group. As a result of these cooperative efforts truly satisfactory comprehensive guarded hot plate and heat flow meter test method documents have been developed and are in the promulgation stage.

During the past decade or so there has been considerable additional worldwide efforts expended in the area of measurement of thermal performance of insulations by these two recognized methods. Efforts have been directed toward better analysis of the methods in order to devise improved means of attainment of one-dimensional heat flow or to obtain more precise corrections. In addition, great improvements have been made in apparatus design and operation while newer concepts have been also developed. There was a need to bring these experiences together especially as C16.30 in the process of undertaking further revisions of the documents to reflect the newer information.

The Symposium was organized, therefore, to provide a forum for an exchange of these ideas and experiences by the group of international workers and to document the current state of the art specifically involved with these two methods. The ultimate goal being to enable better measurements to be made on all types of insulations at all temperatures and conditions with comparable reliability whoever is involved and whichever of these two techniques are utilized.

The international group of papers covers a wide variety of subjects from basic analyses, apparatus design and development, and instrumentation details to applications and use. These papers should provide the required stimulus to achieve the desired goal.

The Chairmen wish to thank all of the authors, session chairmen, and many reviewers for their efforts in making the Symposium a success. Due to illness our good friend and colleague Dr. Karl Heinz Bode was unable to present his paper but with the approval of Pergammon Press we have included an important contribution of his which is directly relevant to this subject. Finally, we wish to thank The National Research Council of Canada for its financial support and particularly Ben Stafford of the Division of Building Research for his stalwart local organizational efforts.

C. J. Shirtliffe

Division of Building Research, National Research Council, Canada K1A 0R6; editor.

R. P. Tye

Thermatest Department, Dynatech R/D Company, Cambridge, MA 02139; editor.

Apparatus Analyses and Error Analyses

Louis R. Troussart[1]

Analysis of Errors in Guarded Hot Plate Measurements as Compiled by the Finite Element Method

REFERENCE: Troussart, L. R., **"Analysis of Errors in Guarded Hot Plate Measurements as Compiled by the Finite Element Method,"** *Guarded Hot Plate and Heat Flow Meter Methodology, ASTM STP 879*, C. J. Shirtliffe and R. P. Tye, Eds., American Society for Testing and Materials, Philadelphia, 1985, pp. 7-28.

ABSTRACT: The object of this research was to study the effect of various design parameters of a guarded hot plate (GHP) on the error of measurement due to gap unbalance without having to resort to simplifying assumptions as has been the case in existing analytical studies. A better knowledge of this effect could lead to better national standards. It could explain some of the discrepancies often found between the results of round-robin tests. The method used is the finite element (FE) method applied to an axisymmetric GHP. The latter choice is justified by the fine discretization of the domain which is achievable without requiring a huge computer memory. This feature allows us to study the influence of thermopile wires crossing the gap.

The results show that for a given gap unbalance, one is not free to arbitrarily choose the thermopile wire diameter, nor the temperature drop across a given specimen thickness, if one seeks accurate measurements. One cannot achieve precise measurements with one apparatus and only one set of operating conditions that would apply to a broad range of thermal conductivities.

The value of the FE method to solve the differential equation of heat conduction while satisfying the operating boundary conditions of the GHP apparatus is also demonstrated.

KEY WORDS: guarded hot plate, errors, finite element, thermopile

Nomenclature

- h Convection heat transfer coefficient, $W/m^2 K$
- r, z Cylindrical coordinates, m
- q Boundary heat flux density, W/m^2
- t Sample thickness, m
- N_i Interpolation function pertaining to node i

[1]Director of Manufacturing and Process Development, Pittsburgh Corning Europe N. V., B-3980 Tessenderlo, Belgium.

S Area, m^2
T Temperature, K
λ Thermal conductivity, W/m K
θ Cylindrical coordinate, radians

Subscripts

b Denotes bottom
e Denotes element
t Denotes top
i Local node index
∞ Used with T to indicate convective-air temperature

In 1981, we had started a three-dimensional study of the guarded hot plate (GHP) by the finite element (FE) method [*1*]. In this early work we used a square GHP model because our laboratory apparatus consisted of square plates. With that model, it was not possible to examine the influence of thermopile wires crossing the gap because of the limitation in computer memory. Yet, we felt that these wires would have a great influence on the error of measurement as a function of the gap temperature unbalance. Therefore, we chose a circular GHP model because it could be generated by one 360° revolution of a half diametral section, and such a model enabled us to use a very fine grid that includes thermopiles. The study is not finished. To date, we have studied insulating specimens of 0.042 W/m K thermal conductivity. We plan to study specimens of various thermal conductivities including 0.21, 1.05, and 0.021 W/m K or so. We may use higher order elements in future work.

Governing Equations

Assuming the steady-state and constant-thermal properties in the domain, the differential equation of heat conduction in cylindrical coordinates for the axisymmetric case when there is no internal heat generation is given by

$$\lambda \frac{\partial^2 T}{\partial r^2} + \frac{1}{r} \lambda \frac{\partial T}{\partial r} + \lambda \frac{\partial^2 T}{\partial z^2} = 0 \tag{1}$$

A half diametral cross section representing the upper part of the apparatus is given Fig. 1 where

S_1 extends from the center of the plate to the outside of the ring and coincides with the cold plate; the latter is a cold sink.

S_2 extends from the center of the center plate to its heater edge. The FE method can account for a heater that does not extend exactly to the edge of the central plate; the heater does in this report.

S_3 designates the vertical cylindrical surface of a glass fiber insulation plus

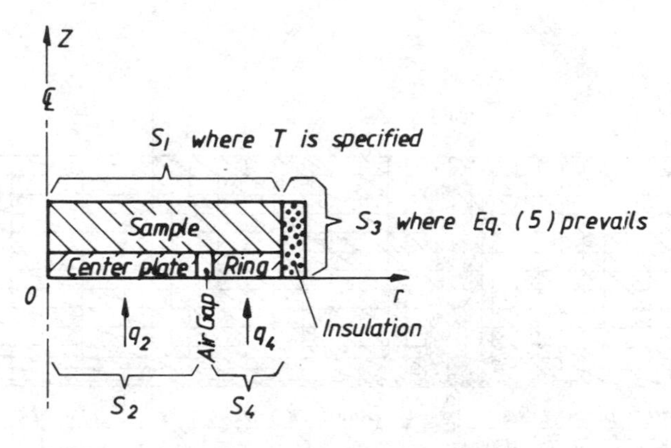

FIG. 1—*Half cross section of top half of circular GHP.*

its ringlike top surface, that is, the cold sink is not assumed to extend over the top of the insulation; a convection boundary condition reigns over S_3.

S_4 extends exactly over the guard ring heater, that is, its width need not be exactly that of the guard ring; it does in this report.

The boundary conditions used in this report are given by

$$T = T_0 \quad \text{on} \quad S_1 \tag{2}$$

$$-\lambda \frac{\partial T}{\partial z} = q_2 \quad \text{on} \quad S_2 \tag{3}$$

$$-\lambda \frac{\partial T}{\partial z} = q_4 \quad \text{on} \quad S_4 \tag{4}$$

where q_2 and q_4 are > 0.

$$\lambda \frac{\partial T}{\partial r} + h(T - T_\infty) = 0 \quad \text{on} \quad S_3 \quad \text{(convection)} \tag{5}$$

Though we assumed that T_0, q_2, and q_4 are radially uniform, it must be emphasized that the FE treatment does not require such an assumption. However, the use of the circular model assumes that these quantities are uniform along the θ-direction.

Finite Element Models

The meshes used in this report consisted of, respectively, 363 nodes with 640 elements and 585 nodes with 1064 linear triangular elements. The latter, used for plates with thermopiles, is shown Fig. 2. It shows that a finer mesh was used

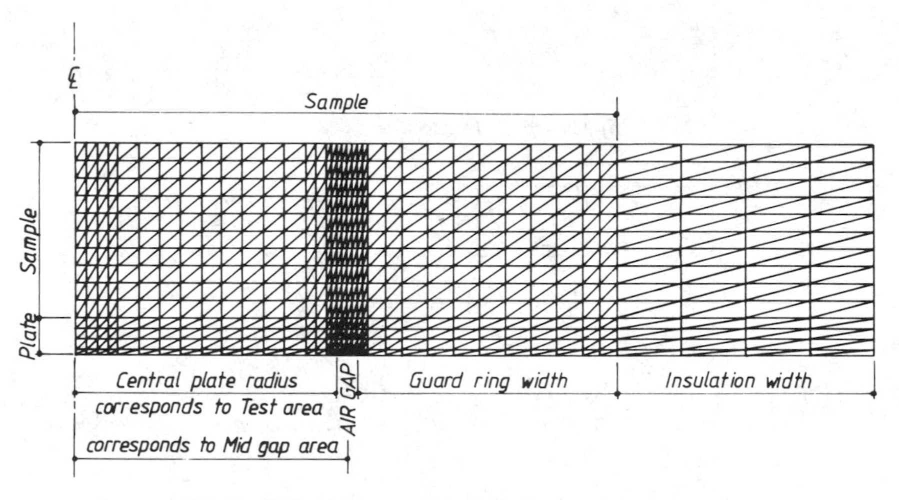

FIG. 2—*Grid used for studying the influence of thermopiles.*

at some locations. A computer program based on the equations yielded by the variational formulation for the heat conduction equation under realistic boundary conditions was written by the author. Changes in the topology or in assumed operating conditions are considered easily by altering a few lines in the program.

The presence of a thermopile is taken care of as follows.

The FEs that correspond to the wires actually generate an annular disk in the model because of the axisymmetry. Radiation across the gap was not considered because it is negligible as compared to conduction through thermopile wires.

Also, natural convection in the air gap was assumed to be negligible, because the latter has very minor temperature gradients along its walls.

The disk is considered to be orthotropic because the thermopile does not influence the heat flow rate perpendicular to the plates, though it does so radially. Its thermal conductivity in the radial direction is calculated from the formula for the steady-state heat transfer in an infinite hollow cylinder, modified as given in the Appendix by assuming that the total heat exchanged between inner and outer cylindrical surfaces of the annular disk is equal to that passing through all the wires evenly distributed around the ringlike gap. Its thermal conductivity in the z-direction is that of the metal plate, and that in the θ-direction can have any value since $dT/d\theta = 0$ (axisymmetry).

The influence of unevenly distributed lead wires cannot be examined with this model.

Advantages of the FE Method

The FE method has many advantages over an exact mathematical solution, given that both assume radial symmetry:

1. Simplifying assumptions need not be made about the gap temperature balance or unbalance, or the gap material, or any complicated though realistic boundary conditions.
2. It allows calculation of the lateral heat flow through the gap as well as the lateral heat flow through the specimen at any location.
3. It can account for nonuniform temperature distribution in the cold plates.
4. It can account for a nonuniform temperature at the outside edge of the specimen or a convection boundary condition at the specimen periphery, whether the latter is insulated or not.
5. The method can consider plate heaters that do not extend to the plate edges or that give a nonuniform heat flux density, radially.
6. The temperatures at as many locations as there are nodes in the model are automatically known from the model solution. So are the heat flows at all its boundaries. For any given specimen, the method gives the gap unbalance that corresponds to any guard ring power input. This helps devising a GHP properly for its expected performance.
7. Last but not least, the method can account for anisotropy in any domain.

All of these features together have not been dealt with in published analytical solutions [*2-6*]. For all of these, simplifying assumptions have been made. For example, one assumes gap temperature equilibrium and neglects the existence of a gap in the actual integration. Or one first assumes that the gap is filled with metal and after integration with a constant λ in the plate region, reintroduces a correction for the actual λ of the gap material. Or one can assume a uniform temperature for the plate and the guard ring plus gap balance instead of different heat flux densities at the central plate and ring, etc.

Of course the FE method is a numerical method. Its accuracy, aside from the assumption of radial symmetry, depends on the fineness of the discretization used and on the order of the elements, that is, the order of the polynomials used for the shape functions (see Appendix).

Definition of Terms Used

Calculated thermal conductivity, in W/m K, is given by

$$\lambda_1 = \frac{P_0 t}{(T_1 - T_0) S_m} \qquad (6)$$

where

P_0 = nominal central plate heater power in W,
t = specimen thickness in m,
T_0 = cold sink temperature in K,
T_1 = weighted average temperature at the bottom surface of the sample, weighted according to the area of elements in contact with the central plate, in K
S_m = midgap area, in m^2, as defined in the Appendix.

Note that the boundary heat flux density q_2 is equal to P_0 divided by the net heater area. It is applied to that part of the boundary covered by the actual heater.

FE-value is the thermal conductivity obtained by

$$\lambda_2 = \frac{Ft}{T_1 - T_0} \tag{7}$$

where F is the average of the heating flux densities going into and coming out from the specimen, in W/m^2. These are weighted over the test area as explained later on.

Gap unbalance is the temperature at the inner side of gap minus the temperature at the outer side of the gap.

Error is defined as the ratio

$$\text{(used value–true value)/true value} \tag{8}$$

where *used value* is λ_1 for the error on calculated lambda and λ_2 for the error on *FE-value*.

Results

For a given specimen conductivity, the computer program puts the following variables into loops with the values indicated:

ΔT across specimen thickness: 5, 10, or 20 K
Assumed true lambda value: 0.042 W/m K
Number of welds for the thermopile: 45 on each side of the gap or none
Wires diameter and their composition: 0.1, 0.3, and 0.5 mm, either copper/constantan or iron/constantan
Central plate radius: 150 and 250 mm
Guard ring width: 150, 50, and 30 mm
Nature of the GHP metal: copper or aluminum
Gap width: 2 and 6 mm
Plate thickness: 3 and 7 mm

Specimen thickness: 2.5, 5, 8, and 10 cm
Insulation width around the apparatus: 150 mm
Convection heat transfer coefficient: 0.05815 W/m^2 K in this report
Ambient air temperature: equal to specimen average temperature (see comments given later on).

Our program uses the ratio of ring power density to central plate power density as a parameter. The output consists mainly of the gap unbalance, the nodal temperatures, among which the hot plate temperature, the heat flux in the specimen elements, the calculated specimen conductivity, the FE-values, and the errors as defined previously.

The convection heat transfer coefficient at the outer insulation surface was set at 0.05815 W/m^2 K, because that value had given a gap unbalance very close to the experimental one for our square GHP.

The influence of the following factors is illustrated more specifically by the figures indicated below and which will be commented afterwards.

Sample thickness, Figs. 3 and 5
Temperature drop on the specimen, Figs. 3 and 5
Air gap width, Figs. 3, 5, and 8
Thermocouple wire alloy, Fig. 6
Thermocouple wire diameter, Fig. 6
Plate metal, Fig. 8
Plate diameter, Fig. 8

The usefulness of the FE method is illustrated Figs. 4 and 7.

A. *Plate Without Thermopile*

The operating conditions are described in Table 1 for Figs. 3 and 4 and in Table 2 for Fig. 5.

Figure 3 gives the calculated thermal conductivity in the function of the gap unbalance. The inset gives the error relative to the true value. It shows that for a given change in gap unbalance, the variation of calculated thermal conductivity is the greatest, the thicker the specimen and the smaller the temperature difference between hot and cold plates. It shows the apparently unexpected result that the calculated λ for a given gap unbalance does not seem to depend on the ring width when the convective-air is at the average specimen temperature and for an apparatus with good lateral insulation (150-mm glass fiber). This can be reconciled with the results obtained from analytical work.

Curve G shows that a wider gap is beneficial since its slope is smaller than that of Curve B. The variation on Curve B is 0.1867 W/m K per degree K unbalance, while for Curve G it is 0.0667, that is, 2.8 times less for a gap 3 times wider.

It must be mentioned that although various ring widths can give the same

TABLE 1—*Conditions common to all curves on Figs. 3 and 4.*

Thermopiles: none
True lambda of specimen: 0.042 W/m K
Test plate radius: 150 mm
Test plate nature: copper
Guarded ring width: 150 mm except as indicated Fig. 3
Width and lambda[a] of insulation: 150 mm and 0.035 W/m K, respectively
Convection air temperature: 273.16 K
Average sample temperature: 273.16 K
Convection heat transfer coefficient: 0.05815 W/m^2 K

Conditions Particular to Some Curves

Curve	Specimen Thickness, cm	Temperature Difference, K	Air Gap, mm
A	10	5	2
B	10	20	2
C	5	5	2
D	5	20	2
E	2.5	5	2
F	2.5	20	2
G	10	20	6

[a]Anistropy of glass fiber insulation was not considered here.

curve for the calculated lambda in the function of the gap unbalance, the latter is not obtained for the same ratio of guard ring power density to central plate power density, all other things equal, and this ratio is not proportional to the guard ring width.

It can be also seen that if a systematic error is made in the measurement of the gap unbalance, one will incorrectly find that the specimen conductivity would vary with its thickness. The thickness effect, if due to this cause, is reversed if unbalance is reversed.

The "error" insert in Fig. 3 shows that the error can be extremely significant for thick specimens at say ± 0.1 K gap unbalance.

Figure 4 gives the FE-value of thermal conductivity in functions of the gap unbalance. The inset gives the error relative to the true value, using an enlarged ordinate scale. It is seen that the FE-value is practically independent of the gap unbalance except for Curve A. The latter was for a very narrow guard ring and a very thick specimen. It seems to indicate that even at gap balance there is a radial heat loss from the specimen within the central plate area.

It should not be construed from the examples given that the analysis is limited to having ambient temperature around the apparatus at the average specimen temperature, which is common practice. Computer runs made with other values for the ambient temperature lead to the same conclusions. These are also valid if the value of h is different from the value chosen here. What does change with these other assumptions is the ratio of guard ring power density to

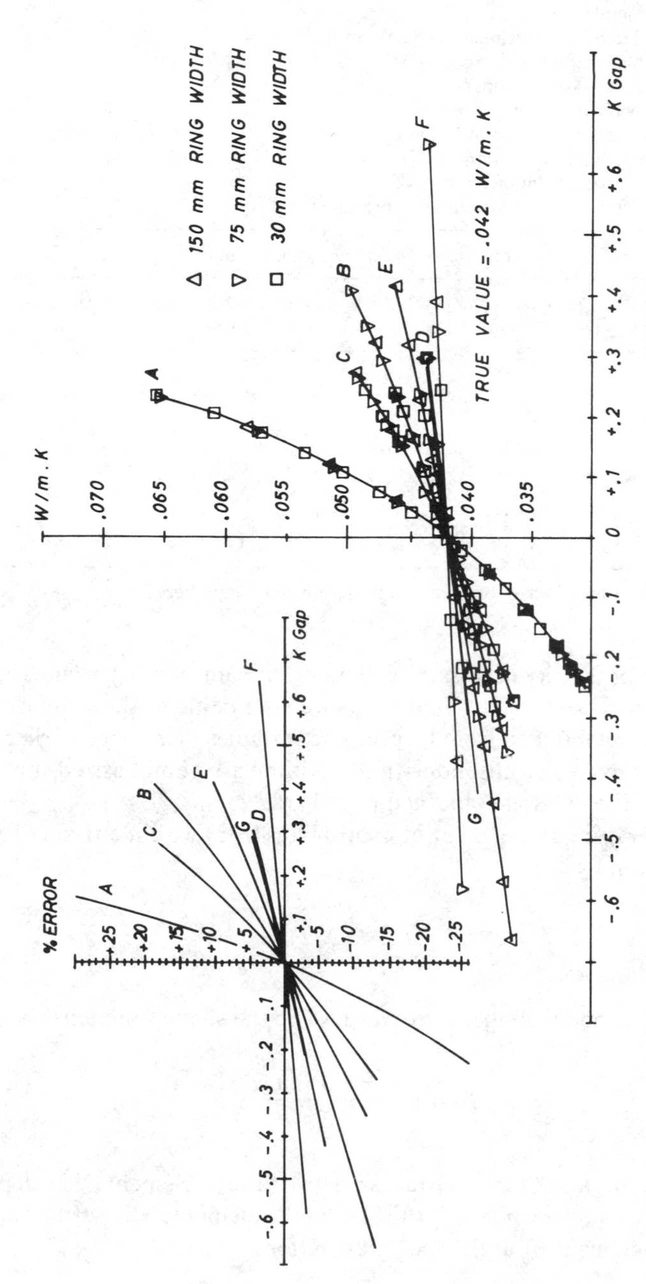

FIG. 3—*Calculated thermal conductivity versus gap temperature difference (Table 1).*

TABLE 2—*Conditions common to all curves on Fig. 5.*

Thermopiles: none
True lambda of specimen: 0.042 W/m K
Test plate radius: 150 mm except for A where it was 30 mm
Test plate nature: copper
Guarded ring width: 30 mm
Width and lambda[a] of insulation: 150 mm and 0.035 W/m K, respectively
Convection air temperature: 273.16 K
Average sample temperature: 273.16 K
Convection heat transfer coefficient: 0.05815 W/m^2 K

Conditions Particular to Some Curves

Curve	Specimen Thickness, cm	Temperature Difference, K	Air Gap, mm
A	8	5	2
B	8	20	2
C	10	5	2
D	10	20	2
E	8	5	6
F	8	20	6
G	10	5	6
H	10	20	6

[a]Anisotropy of glass fiber insulation was not considered here.

test plate power density necessary to achieve the same gap unbalance as in the example given. Due to space limitations we are unable to show the numerous curves that were obtained from looping the variables mentioned earlier.

In order to avoid any misunderstanding it must be emphasized that the results for the FE-value were obtained as follows:

The temperature at any point of coordinates (r, z) within a triangular linear element is given by

$$T = \sum_{1}^{3} N_i T_i \tag{9}$$

where T_i are the nodal temperatures and N_i are the shape functions given by

$$N_i = \frac{A_i + B_i r + C_i z}{2S_e} \tag{10}$$

and where A_i, B_i, and C_i are constants relative to each element (they depend on its nodal coordinates), while S_e is the area of the element. The vertical heat flux density in an element E_b at the specimen bottom surface is

$$-\lambda\left(\frac{\partial T}{\partial z}\right)_{E_b} = -\lambda\left(\sum_{1}^{3} \frac{\partial N_i}{\partial z} T_i\right)_{E_b} = -\lambda\left(\sum_{1}^{3} \frac{C_i T_i}{2S_e}\right)_{E_b} \tag{11}$$

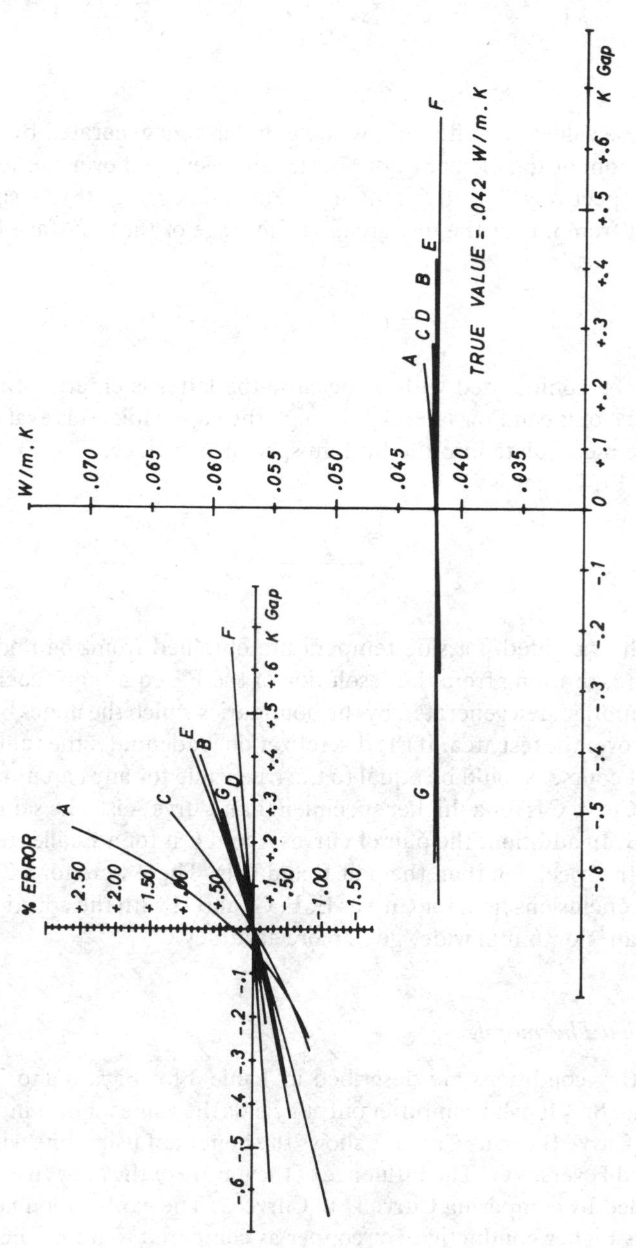

FIG. 4—*FE-values versus gap temperature difference (Table 1).*

The vertical heat flux in an element E_t at the specimen top surface is

$$-\lambda\left(\frac{\partial T}{\partial z}\right)_{E_t} = -\lambda\left(\sum_1^3 \frac{\partial N_i}{\partial z} T_i\right)_{E_t} = -\lambda\left(\sum_1^3 \frac{C_i T_i}{2S_e}\right)_{E_t} \quad (12)$$

Each of these values is multiplied by the annular area generated by, respectively, the bottom or top element boundaries and weighted over the test area. This yields, respectively, F_1 and F_2, that is, the heat flux going into the specimen and going out from it over the test area. The average of these surface fluxes is then

$$F = (F_1 + F_2)/2$$

F_1 may not be confounded with q_2 because the latter is entering the metal plate, and some of it can flow laterally through the gap, while F_1 is evaluated at the exit of the metal plate into the bottom specimen surface. The FE-value is then given by Eq 7, that is

$$\lambda_2 = \frac{Ft}{T_1 - T_0}$$

where T_1 is the weighted hot side temperature obtained from the nodal temperatures there, resulting from the resolution of the FE equations, each multiplied by the annular area generated by the boundaries which the nodes belong to and weighted over the test area. If the discretization is adequate, the value found using Eq 7, of course, should be equal to the true value for any gap unbalance.

On Fig. 5, Curve C is for a thicker specimen than Curve A is. The same is true for D versus B. In addition, the pair of curves A and C is for a smaller temperature drop on the specimen than the pair B and D is. These were for a 2-mm air gap. Similar conclusions hold for Curves E, F, G, and H with the additional one that the air gap, now 6 mm wide, gives more accuracy.

B. *Plate With a Thermopile*

The operating conditions are described in Table 3 for Figs. 6 and 7 and in Table 4 for Fig. 8. A typical computer output yields the values shown in Table 5.

On Fig. 6, Curve B versus Curve A shows the benefit of using fine wires, also E versus D and H versus G. The influence of the nature of the alloys used for the wires is revealed by comparing Curve D to Curve C. The explanation lies in the nearly 5 times higher conductivity of copper as compared to iron. The thicker the wires diameter, the more one looses the advantage of having a thermopile made from many thermocouples.

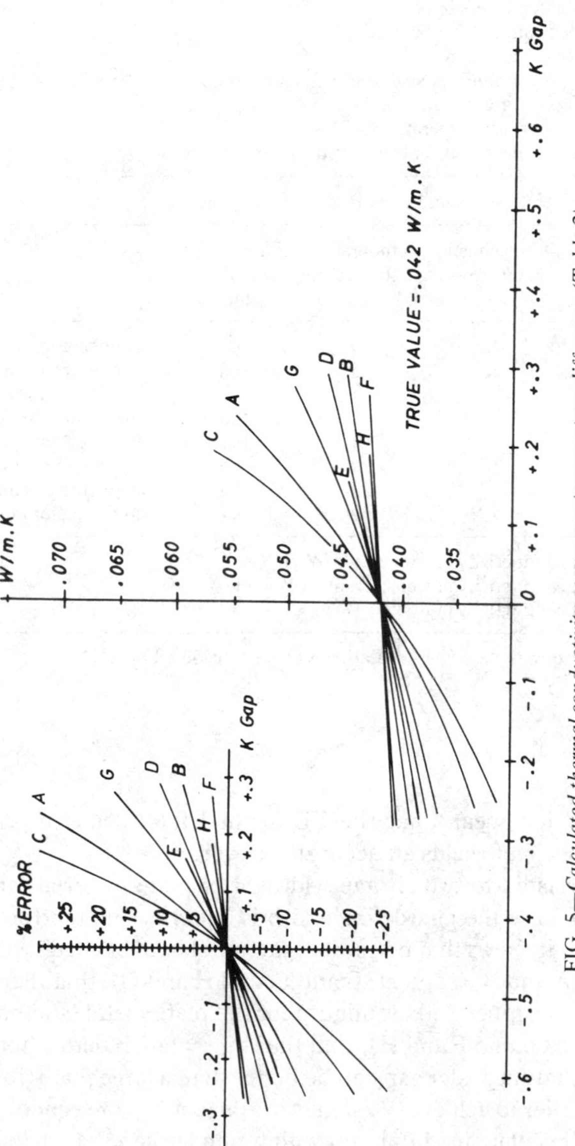

FIG. 5—*Calculated thermal conductivity versus gap temperature difference (Table 2).*

TABLE 3—*Conditions common to all curves on Figs. 6 and 7.*

Thermopile: 45 junctions at each side of annular gap
True lambda of specimen: 0.042 W/m K
Test plate radius: 150 mm
Test plate nature: copper
Guarded ring width: 150 mm
Air gap width: 2 mm
Width and lambda[a] of insulation: 150 mm and 0.035 W/m K, respectively
Convection air temperature: 273.16 K
Average sample temperature: 273.16 K
Convection heat transfer coefficient: 0.05815 W/m^2 K

Conditions Particular to Some Curves

Curve	Specimen Thickness, cm	Temperature Difference, K	Thermopile Diameter, mm	Wire Type
A	2.5	5	0.5	copper/constantan
B	2.5	5	0.3	copper/constantan
C	5	5	0.3	copper/constantan
D	5	5	0.3	iron/constantan
E	5	5	0.1	iron/constantan
F	2.5	20	0.5	copper/constantan
G	5	20	0.3	copper/constantan
H	5	20	0.1	iron/constantan

Thermal conductivity of copper 393 W/m K
Thermal conductivity of Constantan 22 W/m K
Thermal conductivity of iron 80 W/m K

[a] Anistropy of glass fiber insulation was not considered here.

From Fig. 7 it appears that the FE method for calculating the heat flux through the specimen yields an accurate answer.

Figure 8 is related to two air gap widths, two plate materials and two plate radii. For each case the guard ring width is equal to one tenth of the central plate diameter. It shows that better measurements will be made with the widest gap (compare A and B, C and D, E and F, and H and G), that there is little difference between copper and aluminum for the plate metal (compare A and C, B and D, E and G, and F and H), that the larger plate is better (for example, A versus E) but that the wider gap can be better than a large plate (for example, F versus A). In order to achieve the same accuracy, it is more economical to build a 150-mm-radius plate in aluminum with 6-mm air gap than a 250-mm-radius plate in copper with 2-mm air gap (H is better than A). That aluminum plate is practically as good as the larger copper plate with 6 mm air gap (H is nearly as good as B).

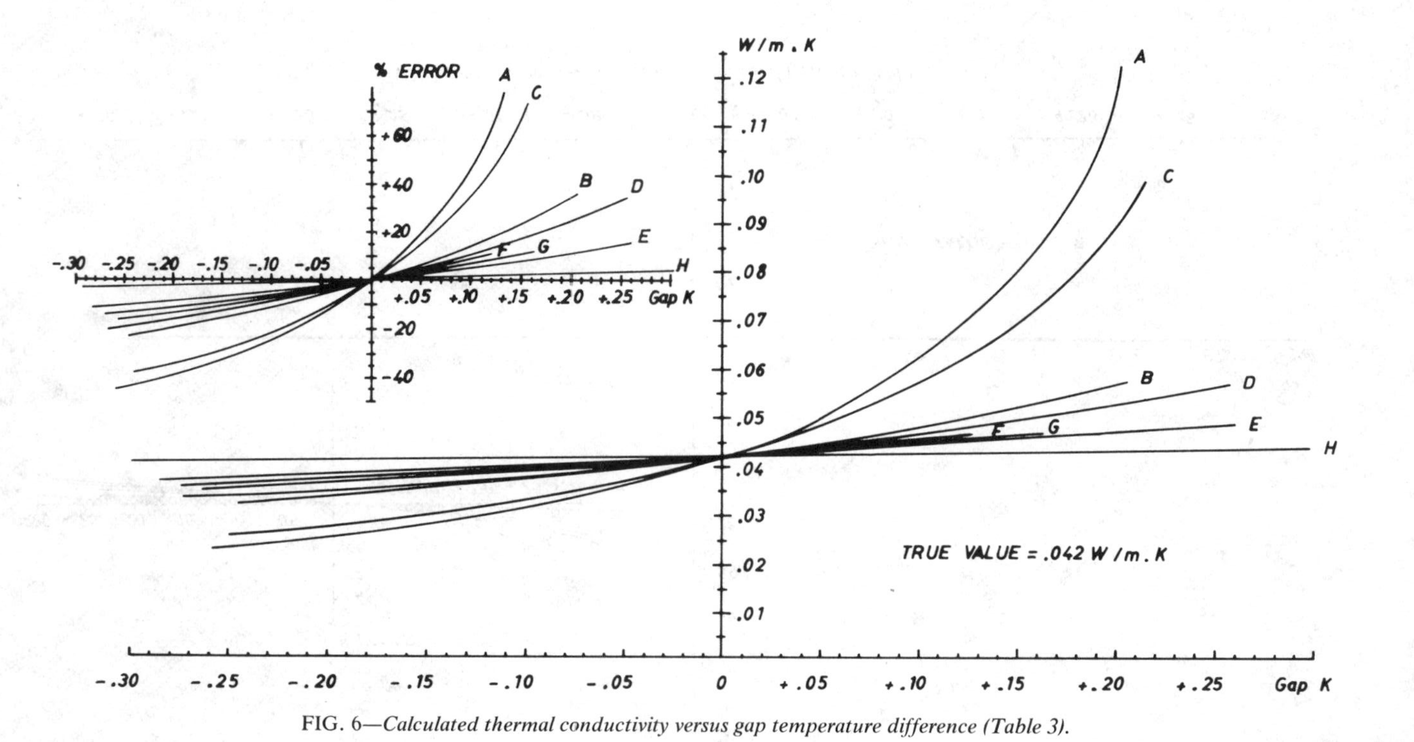

FIG. 6—*Calculated thermal conductivity versus gap temperature difference (Table 3).*

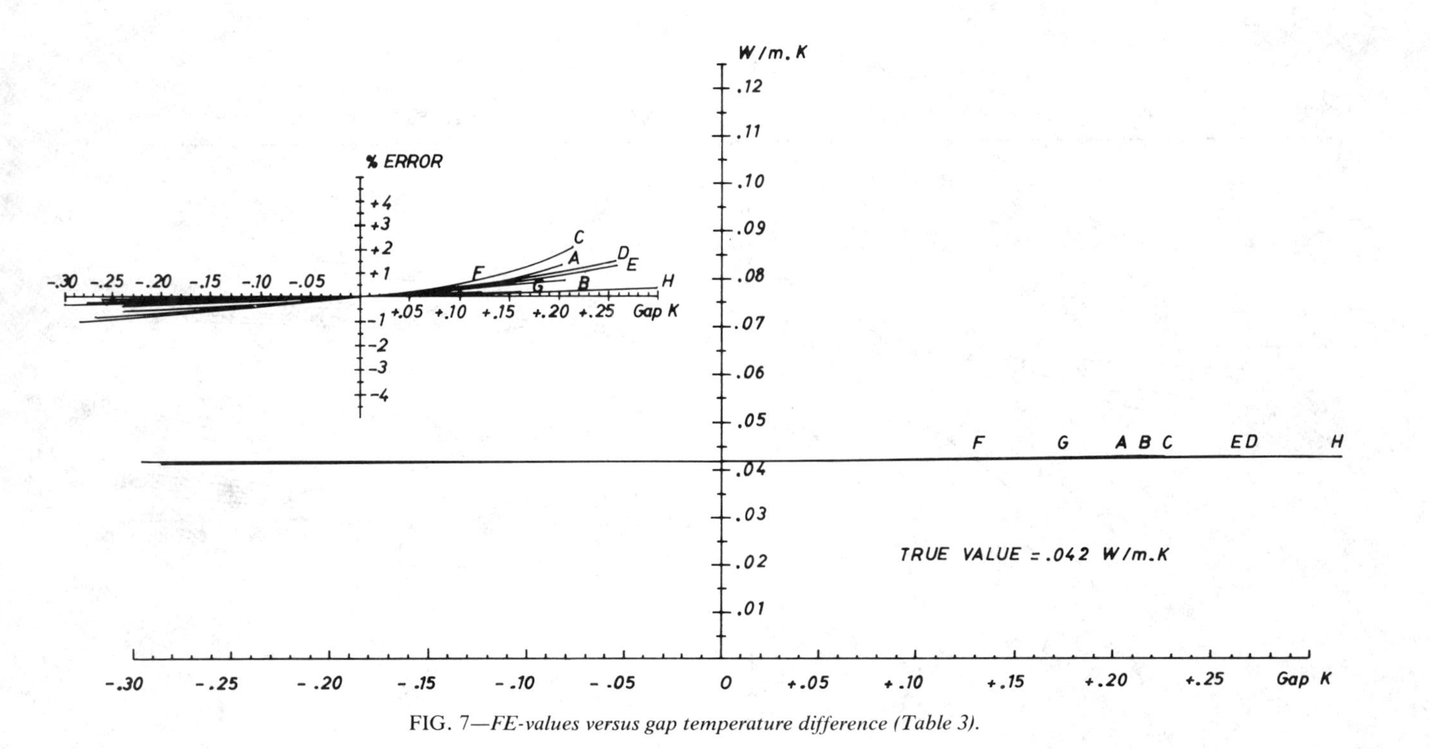

FIG. 7—*FE-values versus gap temperature difference (Table 3).*

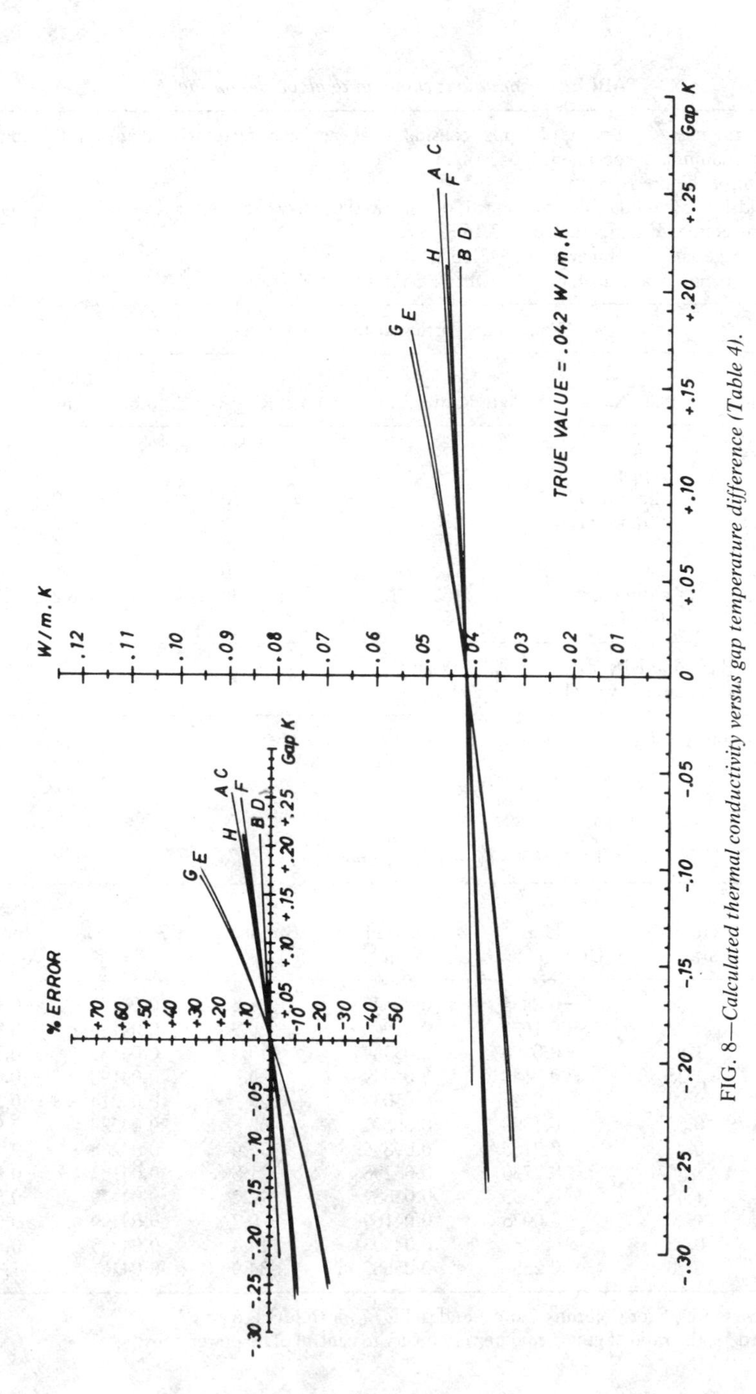

FIG. 8—*Calculated thermal conductivity versus gap temperature difference (Table 4).*

TABLE 4—*Conditions common to all curves on Fig. 8.*

Thermopile: 45 junctions of copper-constantan at each side of gap; wires diameter: 0.3 mm
True lambda of specimen: 0.042 W/m K
Sample thickness: 5 cm
Width and lambda[a] of insulation: 150 mm and 0.035 W/m K, respectively
Convection air temperature: 273.16 K
Average sample temperature: 273.16 K
Convection heat transfer coefficient: 0.05815 W/m^2 K

Conditions Particular to Some Curves

Curve	Plate Nature	Plate Radius, mm	Guard Ring Width, mm	Air Gap, mm
A	copper	250	50	2
B	copper	250	50	6
C	aluminum	250	50	2
D	aluminum	250	50	6
E	copper	150	30	2
F	copper	150	30	6
G	aluminum	150	30	2
H	aluminum	150	30	6

Thermal conductivity of copper 393.0 W/m K
Thermal conductivity of aluminum 209.3 W/m K

[a]Anisotropy of glass fiber insulation was not considered here.

TABLE 5—*Typical computer output values.*[a]

Curve	Guard[b] Parameter	Gap Unbalance, K	Calculated λ_1, W/m K	Error on λ_1, %	FE-Value λ_2, W/m K	Error on λ_2, %
C	2.30	−0.2435	0.02547	−39.4	0.04175	−0.60
C	1.95	−0.1670	0.02906	−30.8	0.04180	−0.47
C	1.60	−0.0905	0.03384	−19.4	0.04188	−0.28
C	1.25	−0.0140	0.04049	−3.6	0.04198	−0.05
C	0.90	0.0625	0.05039	−20	0.04214	0.33
C	0.55	0.1390	0.06672	58.8	0.04239	0.93
C	0.20	0.2155	0.09896	135.6	0.04288	2.10
D	1.70	−0.2730	0.03298	−21.5	0.04163	−0.88
D	1.45	−0.1405	0.03682	−12.3	0.04179	−0.50
D	1.20	−0.0075	0.04169	−0.7	0.04199	−0.02
D	0.95	0.125	0.04803	14.4	0.04225	0.60
D	0.70	0.258	0.05665	34.9	0.04261	1.45

[a]Relative to Fig. 6 for Columns 4 and 5 and to Fig. 7 for Columns 6 and 7.
[b]"Guard" is the ratio of guard ring power density to central plate power density.

C. *Conductivity Derived from a Heat Balance*

The FE method yields accurate values for the various heat fluxes in a GHP even for gap unbalance conditions. This fact suggests to calculate the λ-value of a specimen from a series of simultaneously recorded temperatures and a heat balance equation for the GHP. Unfortunately, in order to do this, one would need to measure temperatures at the location of the nodal points of the boundary elements. This is not possible at the inside points of the elements.

We attempted to turn around the difficulty by assuming that the radial and vertical temperature distributions were linear. This yielded the calculated lambda with 3% error when gap unbalance was 0.236 K and with 0.3% error for 0.0018 K unbalance. Therefore, such a heat balance method offers no advantage over the classical procedure.

D. *Reliability of Interpolation Practice*

Under certain operating conditions the variation of the calculated thermal conductivity with gap unbalance is not linear. In that case, the curve is always concave towards the direction of increasing thermal conductivity. As a consequence, the practice, often encountered to save time, consisting of taking two experimental readings, respectively, with a negative and positive gap temperature difference, and then interpolate at zero gap differential, is unreliable. Prior to using such practice it should be checked that the variation in question is linear within the range of gap unbalance considered.

Comparison With Recent Analytical Work by Others

There is no contradiction between the results in this work and those reported by Bode [*6*] for the influence of the guard ring width on GHP measurements.

Indeed, Dr. Bode found that the error on the k-value within the frame of his assumed boundary conditions was at gap balance

$$\frac{\lambda_m - \lambda_r}{\lambda_r} = f_1 + \frac{T_{av} - T_{amb}}{\Delta T} \times f_2 \tag{13}$$

where

λ_m = measured value,
λ_r = true value,
T_{av} = the average of hot and cold face temperatures,
T_{amb} = outside convective-air temperature, and
f_1, f_2 = algebraic expressions derived from his analytical integration and depend mainly on ring width.

He showed that f_1 is negligible.

Now, if $T_{av} = T_{amb}$ as is the case in our work reported here, Eq 14 would also give

$$(\lambda_m - \lambda_r)/\lambda_r = f_1 \cong 0$$

regardless of the ring width.

There is no doubt whatsoever that if the FE method were applied using Dr. Bode's boundary conditions, it would yield the same results as his. We recall that those conditions were a uniform temperature at the cold and hot sides from dead center to the outside of the ring, thus assuming gap balance and the convection heat transfer at the uninsulated periphery. In actual practice the ring temperature at its outside edge is not equal to that of its inner edge. In addition, the actual boundary conditions in the operating practice are a constant heat flux density in the central plate and another heat flux density in the ring, which are the ones that we considered in our analysis.

Conclusion

More systematic computer runs will be necessary in order to obtain detailed recommendations for the GHP parameters. These will depend on the range of k-values it is built for, on the thickness of the specimens and on the wanted accuracy in the measurement of thermal conductivity. Neither the number of thermal junctions constituting the thermopile nor the wire diameter may be chosen at will. Too many junctions, especially with wires diameter over 0.2 mm, will result in steeper variation of the hot plate temperature with gap unbalance and part, if not all, of the benefit of using a thermopile will be lost. The allowance, often found in national standards, to take actual readings when the gap is in balance at ± 0.1 K seems to be inadequate, at least with specimens thicker than 2.5 cm and temperature drop below 20 K.

We believe that when measuring the λ-value of good insulating materials (0.045 W/m K and below) in thicknesses over 2.5 cm, one should use at least 20 K temperature drop across the specimens and no thermopile wires should be of a diameter larger than 0.2 mm. At present, the FE method seems to be a useful way of studying the influence of various parameters on the GHP measurements for realistic boundary conditions. We feel that the FE method is better suited than the finite differences method for problems with irregular boundaries and nonhomogeneous properties, though both methods converge to the same values when the number of nodes is increased. Other authors [*7–10*] seem to agree to this. Huebner [*10*] mentions especially the work of Emery and Carson [*11*], but the latter was not available to the author.

APPENDIX

1. Annular Disk Radial Conductivity

We consider the upper (or lower) half of a GHP sliced through the middle of its total thickness.

Let N be the number of thermocouple junctions at each side of gap, that is, there are two N-junctions evenly distributed on two π-radians for the sliced half; R_1 be the inner radius, that is, the central plate radius, m; R_2 be the outer radius, m; G be the gap width, m; λ_{w1} be the thermal conductivity of one of the thermocouple wires, W/m K; λ_{w2} be that of the other one, W/m K; Φ be the wire diameter, m; and d be the disk thickness which depends on the FE discretization used, m.

One can easily show that in order for the total radial heat flow through the annular disk to be equal to that going through the two N-wires of length equal to G, the annular disk radial conductivity is given by

$$\lambda_r = N \times (\lambda_{w1} + \lambda_{w2}) \times \Phi^2 \times \ln \frac{R_2}{R_1} / 8Gd$$

2. Influence of the Order of the Elements

In this report, linear triangular elements were used. The number of elements can be reduced by using higher order elements. Lately, we made a program using quadratic triangles. The main difference at this point between the two types is that in linear elements the heat flux is constant within the element, while in quadratic elements the heat flux varies linearly within the element. Computer runs using, respectively, 1064 linear triangles or 168 quadratic triangles compare as follows: Computing time is somewhat longer for the latter scheme because of the higher order of integration in obtaining the coefficients of the global conduction matrix, but it is very accurate though it uses less elements. For a gap unbalance of 0.268 K the quadratic elements gave a calculated λ-value equal to 0.03097, and for the same gap unbalance, the linear elements gave the λ-value 0.03093 W/m K. However, what does change with either the number of the elements or with their order, is the curvature of the curves. This means that when studying the influence of various conditions on the calculated λ-value one must compare curves obtained with the same number and order of elements.

3. Midgap Area

With the symbols used in item 1, we define midgap area as that of a circle of radius

$$R = [(R_1^2 + R_2^2)/2]^{1/2}$$

For the examples which we have used in this report, using the average radius of value $(R_1 + R_2)/2$ is practically equivalent as the following shows: For $R_1 = 150$ mm and $R_2 = 152$ mm one obtains, respectively, radii values of 151.003 and 151. For $R_1 = 150$ mm and $R_2 = 156$ mm those values become 153.029 and 153. The relative differences in areas are, respectively, 0.0043 and 0.038%.

References

[1] Troussart, L. R., *Journal of Thermal Insulation*, Vol. 4, 1981, pp. 225-254.
[2] Donaldson, I. G., *Quarterly Applied Mathematics*, No. 19, 1961, pp. 205-219.
[3] Donaldson, I. G., *British Journal of Applied Physics*, Vol. 13, 1962, pp. 598-602.

[4] Woodside, W., *Review of Scientific Instruments*, Vol. 28, No. 12, 1957, pp. 1033–1037.
[5] Somers, E. V. and Cyphers, J. A., *Review of Scientific Instruments*, Vol. 22, No. 8, 1951, pp. 583–586.
[6] Bode, K. H., *International Journal of Heat Mass Transfer*, Vol. 23, 1980, pp. 961–970.
[7] Desai, C. S. and Abel, J. F., *Introduction to the Finite Element Method*, Van Nostrand-Reinhold, New York, 1972, pp. 4–5 and pp. 435–438.
[8] Myers, G. E., *Analytical Methods in Conduction Heat Transfer*, MacGraw Hill, New York, 1971, p. 342 and pp. 385–387.
[9] Gallagher, R. H., *Finite Element Analysis Fundamentals*, Prentice Hall, Englewood Cliffs, NJ, 1975, pp. 71–74.
[10] Huebner, K. H., *The Finite Element Method for Engineers*, Wiley Interscience, New York, 1975, p. 253.
[11] Emery, A. F. and Carson, W. W., "An Evaluation of the Use of the Finite Element Method in the Computation of Temperature," Sandia Lab Report SCL-RR-69-83, Aug. 1969, ASME Paper 69 WA/HT-38.

Karl-Heinz Bode[1]

Thermal Conductivity Measurements with the Plate Apparatus: Influence of the Guard Ring Width on the Accuracy of Measurements*

REFERENCE: Bode, K.-H., **"Thermal Conductivity Measurements with the Plate Apparatus: Influence of the Guard Ring Width on the Accuracy of Measurements,"** *Guarded Hot Plate and Heat Flow Meter Methodology, ASTM STP 879*, C. J. Shirtliffe and R. P. Tye, Eds., American Society for Testing and Materials, Philadelphia, 1985, pp. 29-48.

ABSTRACT: The temperature field in circular and square plate test specimens is calculated and thereby the error from the heat exchange at the outer edge of the guard ring derived for single and double plate devices under conditions of isothermal hot and cold plates and ideal thermal contact. The error is shown to be proportional to the reduced temperature difference (difference between the mean specimen and the ambient temperatures divided by the difference between the hot and cold plate temperatures) and the margin of error, therefore, can be optimally reduced by an appropriate adjustment in the ambient temperature. Experiments on an extruded foam confirm the results of the analytical calculations. Diagrams and tables which are developed on the basis of the above derived relationships, make possible estimation of margin of error for different measuring arrangements for plate devices of usual sizes.

KEY WORDS: circular plate, square plate, guard ring, double hot plate, measurements

Nomenclature

- A Measuring area
- A_n N^{th} coefficient of the Fourier series (3*a*)
- a Half the edge length of the square plate
- α Heat transfer coefficient

*The original article was printed in German in *International Journal of Heat Mass Transfer*, Vol. 23, 1980, pp. 961–970, and then translated by Language Services at Knoxville, Tennessee. Permissions to reprint have been granted by Pergamon Press, Oak Ridge National Laboratory, and author, K.-H. Bode.

[1]Physikalisch-Technische Bundesanstalt, Bundesallee 100, D-3300 Braunschweig, West Germany.

$\beta_{n,m}$ Abbreviation according to Eq 8
$C_{n,m}$ Coefficient of the double series (Eq 3*b*)
c Half the outer edge length of the square guard ring
θ Temperature, ϑ
θ_1 Isothermal cold plate temperature
θ_2 Isothermal hot plate and guard ring temperature
θ_m Arithmetic mean temperature $0.5\ (\theta_1 + \theta_2)$
θ_u Constant ambient temperature
$\Delta\theta$ Temperature difference $(\theta_2 - \theta_1)$ between hot and cold plates
I_0, I_1 Modified Bessel functions of the first type and of the zero or first order
F_1, F_2 Error factors Eq 9
λ Thermal conductivity of the specimen
λ_s Apparent, measured thermal conductivity of the specimen
λ_D Thermal conductivity of the insulating material
p Dimensionless heat exchange parameter Eqs 5*a* or 5*b*
r, z Cylindrical coordinates
r_0 Specimen radius
r_2 Outside guard ring radius
ρ Abbreviation $n\pi r/z_0$
W Heat output
VS Dimensionless ratio of the specimen height z_0 to the plate diameter $2\ r_0$ or the edge length $2a$
VB Dimensionless ratio of the guard ring width $(r_2 - r_0)$ or $(c - a)$ to the plate diameter $2\ r_0$ or to the edge length $2a$
x, y, z Cartesian coordinates
Δx Distance between the outside edge of the guard ring to the "environment"
z_0 Specimen height
ξ_m Mth root of Eq 7

The "guard ring principle," which is used with single and double plate devices for measuring the thermal conductivity of heat insulating materials, became known in the German-speaking area in 1912 through the work of R. Poensgen [*1*].[2] The guard ring is intended to prevent deviations of the temper-

[2] Poensgen mentions two precursors who both reported on guard ring applications at the 2nd International Cold Congress in Vienna in 1910 (Vol. II of the report): E. R. Metz and A. Behne, "Neue Apparate zur Bestimmung des Wärmeleit fähigkeits Koeffizienten" [New Apparatus for the Determination of the Thermal Conductivity Coefficient], p. 193; R. Biquard, "L'effacacité des divers moyens d'isolement thermique des locaux frigorifiques: Essays sur la conductibilité colorifique" [The Effectiveness of Different Methods of Thermal Insulation on Refrigerator Areas: Experiments on Colorific Conductivity], p. 187. A. Berget seems to have been the first to have used a guard ring (or guard cylinder) for thermal conductivity measurements (of mercury) [C.r. 105 (1887), p. 224; *Journal of Physics*, Vol. 2, No. 7, 1888, p. 503]. Berget refers in his articles to the electrical model of W. Thomson, who used a "guard ring" to safeguard a homoge-

ature field in the test piece from the ideal homogeneous field, which is obtained as the solution of the one-dimensional thermal conductivity differential equation, and thereby guarantee the validity of the simple evaluation of the test data.

Qualitatively the following is true: the wider the guard ring is in comparison with the specimen diameter and specimen height, the more it "protects" the temperature field from distortions. Only little is known about the quantitative extent of the protective effect. For example, the dimensioning recommendations of standards both in Germany and abroad (for example, DIN 52612 [*3*]) are based primarily on experimental experience. Several American authors (for example, Woodside [*2*]) have published calculations in which the problem is treated two-dimensionally and nonrealizable boundary conditions are presupposed.

In order to be able to participate in a ring experiment, in which the thermal conductivity of an extruded foam (thermal conductivity close to that of air) was to be measured, it was necessary to have a more accurate knowledge of the quantitative dependence of the boundary error on geometric and thermal quantities of the test setup: while other institutes used devices for square specimen plates with an edge length of 500 mm and a specimen height of 50 mm, the only device available at the Physikalisch-Technische Bundesanstalt (PTB) was a device for circular plates with a diameter of 120 mm. For the PTB apparatus the specimen height would have had to be reduced to a suitable dimension, which was given by empirical values. However, that was not necessarily possible since in the case of foams of sufficiently low bulk density, the measured "effective" thermal conductivity can vary with the height of the specimen due to the effects of radiation.

Therefore the attempt was made to analyze the effect of guard ring width in devices with circular or square plates. The calculations given in the present paper yield a simple law for the boundary error as a function of given quantities for the devices.

The Differential Equations Along with Boundary Conditions and Their Solutions

For the following analyses, the problem is limited by presupposing a homogeneous and isotropic specimen with constant thermal conductivity λ. In accordance with well tried measuring practice for material with relative small λ, the test specimen is to cover the measuring cross-section and the guard ring area to its outer limit without interruption. So that the guard ring effect can

---

neous electric potential field and is probably the father of the guard ring. The oldest source known to the author is Thomson's "Report on Electrometers and Electrostatic Measurements" in Report of the 37th Meeting of the British Association for the Advancement of Science, held at Dundee in September 1867 (1868), London.

be examined alone, other possible errors are eliminated by assuming that the two ends of the specimen including the guard ring are isothermal and the thermal contact between hot or cold plates and the specimen surfaces are complete (assumptions which are realistic given a small λ of the specimen and the high thermal conductivity of the adjacent material—for example, of copper).

Determination of the coordinates: for the circular and square plates, the origin of coordinates is placed in the center of the isothermal cold plate side with temperature θ_1 (see Fig. 1); the axial coordinate z is perpendicular to this isothermal, plane surface; the heating surface, which is also isothermal, is located at z_0 (z_0 is the specimen height). The radial coordinate r applies to the circular plate, with r_0 as the specimen radius and r_2 as the outer limit. The Cartesian coordinates x, y are valid for the square plate, whereby half the edge length is $x = a$ and $y = a$, and the outer limit of the continuous specimen body is $x = c$ and $y = c$.[3] (r_0 or a are lengths to the midpoint between the hot plate edge and the inner edge of the guard ring.) Therefore, πr_0^2 or $4a^2$ is to be inserted as the measuring Surface A. The adjacent "environment" of constant temperature θ_u is the room air when an unprotected test setup is used, and when protective covers are used it is the cover wall at the point $r_2 + \Delta x$ or $c + \Delta x$. For the heat exchange on the outer specimen body edge r_2 or c, Newton's cooling law is valid with the heat transfer coefficient α, which, when a thermal insulating material of thermal conductivity λ_D is used between the specimen wall and the protective cover, is approximately replaced by

$$\alpha \approx \lambda_D \Big/ \left[r_2 \cdot \ln \left(1 + \frac{\Delta x}{r_2} \right) \right]$$

in the case of a circle, or $\alpha = \lambda_D / \Delta x$ in the case of a square.

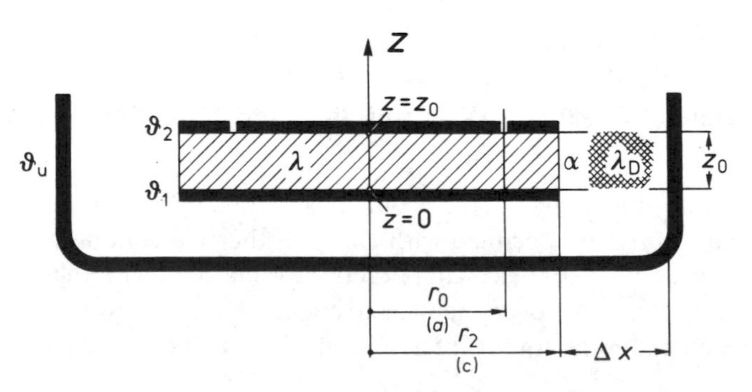

FIG. 1—*Diagram of a plate apparatus with guard ring. Radial coordinates* r *for circular plates (plate diameter* $= 2r_0$*); coordinate values in parentheses for square plates (edge length:* 2a*). Ends of the test body including the guard ring zone (to* r_2 *or* c*) isothermal* (θ_1; θ_2)*. Constant "ambient" temperature* θ_u.

[3] r_1 or b are reserved for later consideration of the configuration having a gap.

Circular Specimen Cross Section

The temperature field $\theta(z, r)$ is determined by the differential equation

$$\frac{\partial^2 \vartheta}{\partial r^2} + \frac{1}{r}\frac{\partial \vartheta}{\partial r} + \frac{\partial^2 \vartheta}{\partial z^2} = 0 \tag{1a}$$

and the boundary conditions

$$\begin{aligned} \vartheta &= \vartheta_1 \quad \text{for} \quad z = 0 \\ \vartheta &= \vartheta_2 \quad \text{for} \quad z = z_0 \\ -\lambda\left(\frac{\partial \vartheta}{\vartheta r}\right)_{r_2} &= \alpha[\vartheta(z, r_2) - \vartheta_u] \end{aligned} \tag{2a}$$

The formula

$$\vartheta(z, r) = \vartheta_1 + (\vartheta_2 - \vartheta_1)\frac{z}{z_0} + \sum_{n=1}^{\infty} A_n \sin\left(\frac{n\pi}{z_0} z\right) I_0\left(\frac{n\pi}{z_0} r\right) \tag{3a}$$

satisfies the differential Eq 1a and also the first two boundary conditions. $I_0(\rho)$ is the modified Bessel function of the first type and zero order.

The constants A_n of the Fourier series are calculated using the last boundary condition and are obtained using a familiar calculation method as

$$A_n = \left.\begin{matrix} -2(\vartheta_m - \vartheta_u) \\ \Delta\vartheta \end{matrix}\right\rangle \cdot \frac{2pz_0}{r_2 n^2 \pi^2 \left[I_1(\rho_2) + \dfrac{p}{\rho_2} I_0(\rho_2)\right]} \left\langle \begin{matrix} \text{for } n = 1, 3, 5\ldots \\ \text{for } n = 2, 4, 6\ldots \end{matrix}\right. \tag{4a}$$

In the above equation the abbreviations are as follows

$$p = \frac{\alpha r_2}{\lambda} \tag{5a}$$

$$\left.\begin{aligned} \vartheta_m &= \frac{\vartheta_2 + \vartheta_1}{2} \\ \Delta\vartheta &= \vartheta_2 - \vartheta_1 \end{aligned}\right\} \tag{6}$$

$\rho_2 = n\pi r_2/z_0$; $I_1(\rho)$ is the modified Bessel function of the first type and first order. When A_n is known, the temperature field $\theta(z, r)$ is known. The Fourier series in Eq 3*a* is convergent; its numerical evaluation poses no difficulty, especially since—depending on claims to accuracy—relatively few terms need to be calculated. The dimensionless heat exchange parameter p according to Eq 5*a* may also become infinite in Eq 4*a*. From Eq 4*a* we then obtain

$$A_n^* = \left.\begin{matrix} -2(\vartheta_m - \vartheta_u) \\ \Delta\vartheta \end{matrix}\right\rangle \cdot \frac{2}{n\pi I_0(\rho_2)} \left\langle\begin{matrix} \text{for } n = 1, 3, 5\ldots \\ \text{for } n = 2, 4, 6\ldots \end{matrix}\right. \tag{4a*}$$

Square Specimen Cross Section

In this case the differential equation reads

$$\frac{\partial^2\vartheta}{\partial x^2} + \frac{\partial^2\vartheta}{\partial y^2} + \frac{\partial^2\vartheta}{\partial z^2} = 0 \tag{1b}$$

and the boundary conditions are

$$\left.\begin{aligned} \vartheta &= \vartheta_1 \quad \text{for} \quad z = 0 \\ \vartheta &= \vartheta_2 \quad \text{for} \quad z = z_0 \end{aligned}\right.$$

and the four equations

$$\left.\begin{aligned} \mp \lambda \left(\frac{\partial\vartheta}{\partial x}\right)_{x=\pm c} &= \alpha[\vartheta(\pm c, y, z) - \vartheta_u] \\ \mp \lambda \left(\frac{\partial\vartheta}{\partial x}\right)_{y=\pm c} &= \alpha[\vartheta(x, \pm c, z) - \vartheta_u] \end{aligned}\right\} \tag{2b}$$

The formulation of the solution is

$$\vartheta(x, y, z) = \vartheta_1 + (\vartheta_2 - \vartheta_1)\frac{z}{z_0} + \sum_{n=1}^{\infty}\sum_{m=1}^{\infty} C_{n,m} \times \sin\left(\frac{n\pi}{z_0}z\right)\left[\cos\left(\xi_m\frac{x}{c}\right)\cosh\left(\beta_{n,m}\frac{y}{c}\right) + \cosh\left(\beta_{n,m}\frac{x}{c}\right)\cos\left(\xi_m\frac{y}{c}\right)\right] \tag{3b}$$

Using

$$p = \frac{\alpha c}{\lambda} \tag{5b}$$

the symbols in Eq 3*b*

$$\xi_m = p \operatorname{ctg}(\xi_m) \tag{7}$$

signify the mth root of this transcendental equation, and

$$\beta_{n,m} = \left[\left(\frac{n\pi}{z_0} c\right)^2 + \xi_m^2\right]^{1/2} \tag{8}$$

The formula Eq 3*b* satisfies the differential equation and also the first two boundary conditions. The remaining four equations (2*b*) all lead to the same conditional equation for the constant $C_{n,m}$, as one can confirm by inserting Eq 3*b* and its corresponding derivations in those equations (2*b*), whereby the relatively simple formula is justified.

The n-part of the double series is again of the Fourier type, whereas the m-part, although also a trigonometric series, has functions of the type

$$\cos\left(\xi_m \frac{x}{c}\right)$$

These cos functions are orthogonal, that is, the following is valid

$$\int_{-c}^{+c} \cos^2\left(\xi_m \frac{x}{c}\right) dx = c\left[1 + \frac{\sin \xi_m \cos \xi_m}{\xi_m}\right]$$

$$\int_{-c}^{+c} \cos\left(\xi_m \frac{x}{c}\right) \cdot \cos\left(\xi_{m'} \frac{x}{c}\right) dx = 0 \quad \text{for} \quad m \neq m'$$

if ξ_m has the meaning of Eq 7. Similar to the laws of a Fourier series, they are suitable for the series representation of arbitrary functions (with the same limitations). For the determination of $C_{n,m}$ in Eq 3*b* the temperature and its derivative (at the respective boundary) are inserted in the four last boundary condition equations (2*b*). Solving [the equations] according to the summarized double series leads in all four cases to the same conditional equation for $C_{n,m}$. One multiplies by

$$\sin\left(\frac{n\pi}{z_0} z\right) \cdot \cos(\xi_m \eta)$$

(where $\eta = x/c$ or $\eta = y/c$) and integrates first over z from $z = 0$ to $z = z_0$, then—paying attention to the given orthogonality relations—over η from $\eta = -1$ to $\eta = +1$. With the abbreviations given in Eq 6, 5b, and 8 and using Eq 7, we obtain

$$C_{n,m} = \frac{-2(\vartheta_m - \vartheta_u)}{\Delta\vartheta} \Bigg\rangle \frac{4}{n\pi} \cdot \frac{\xi_m \sin \xi_m}{[\xi_m^2 + p \cos^2 \xi_m]}$$

$$\times \frac{p}{\beta_{n,m}^2 \left[\sinh \beta_{n,m} + \frac{p}{\beta_{n,m}} \cosh \beta_{n,m} \right]} \Bigg\langle \begin{array}{l} \text{for } n = 1, 3, 5 \ldots \\ \text{for } n = 2, 4, 6 \ldots \end{array} \qquad (4b)$$

In this case as well the limiting process $p \rightarrow \infty$ is possible; Eq 4b then reads

$$C^*_{n,m} = \frac{-2(\vartheta_m - \vartheta_u)}{\Delta\vartheta} \Bigg\rangle \frac{4}{n\pi} \frac{(-1)^m}{\xi_m}$$

$$\times \frac{}{\beta_{n,m} \cosh \beta_{n,m}} \Bigg\langle \begin{array}{l} \text{for } n = 1, 3, 5 \ldots \\ \text{for } n = 2, 4, 6 \ldots \end{array} \qquad (4b^*)$$

Derivation of the Error

Error Expression for the Circular Plate

The thermal output W (which must eventually be corrected for losses through heat dissipation due to the wiring, support elements, etc.), which enters the specimen through the end (z_0, r_0) is

$$W = 2\pi\lambda \int_0^{r_0} \left(\frac{\partial \vartheta}{\partial z} \right)_{z_0} r \, dr = \lambda \left[\pi r_0^2 \frac{\Delta\vartheta}{z_0} + 2\pi r_0 \sum_{n=1}^{\infty} (-1)^n A_n I_1(\rho_0) \right]$$

with the abbreviation $\rho_0 = n\pi r_0/z_0$. With the customary noncritical test evaluation the measured, apparent thermal conductivity is λ_s, and the following equation is valid

$$W = \lambda_s \pi r_0^2 \frac{\Delta\vartheta}{z_0}$$

After equating the two expressions and rearranging, we obtain

$$\frac{\lambda_s - \lambda}{\lambda} = \frac{2z_0}{r_0 \Delta\vartheta} \sum_{n=1}^{\infty} (-1)^n A_n I_1(\rho_0)$$

the error which exists when the boundary effect is not considered. If we insert A_n according to Eq 4a, we obtain finally

$$\frac{\lambda_s - \lambda}{\lambda} = F_1 + \frac{\vartheta_m - \vartheta_u}{\Delta\vartheta} \cdot F_2 \tag{9a}$$

The factors in the above equation, which are only a function of the dimensionless parameter p (Eq 5c) and geometric quantities, are

$$F_1 = \frac{4z_0^2 p}{\pi^2 r_0 r_2} \sum_{n=2,4,6\ldots}^{\infty} \frac{I_1(\rho_0)}{n^2\left[I_1(\rho_2) + \frac{p}{\rho_2} I_0(\rho_2)\right]} \tag{10a}$$

$$F_2 = \frac{8z_0^2 p}{\pi_0^2 r_2} \sum_{n=1,3,5\ldots}^{\infty} \frac{I_1(\rho_0)}{n^2\left[I_1(\rho_2) + \frac{p}{\rho_2} I_0(\rho_2)\right]} \tag{11a}$$

In the limiting case $p \to \infty$ the factors are

$$F_1^* = \frac{4z_0}{\pi r_0} \sum_{n=2,4,6\ldots}^{\infty} \frac{I_1(\rho_0)}{nI_0(\rho_2)} \tag{10a*}$$

$$F_2^* = \frac{8z_0}{\pi r_0} \sum_{n=1,3,5\ldots}^{\infty} \frac{I_1(\rho_0)}{nI_0(\rho_2)}. \tag{11a*}$$

Error Expression for the Square Plate

Similarly the calculations for the square plate yield

$$W = \lambda \int_{-a}^{+a} \int_{-a}^{+a} \left(\frac{\partial\vartheta}{\partial z}\right)_{z0} dx\, dy$$

$$= \lambda\left[4a^2 \frac{\Delta\vartheta}{z_0} + \frac{8c^2\pi}{z_0} \sum_{n=1}^{\infty} \sum_{m=1}^{\infty} (-1)^n C_{n,m} \frac{n}{\xi_m \beta_{n,m}} \times \sin\left(\xi_m \frac{a}{c}\right) \sinh\left(\beta_{n,m} \frac{a}{c}\right)\right]$$

and

$$W = \lambda_s 4a^2 \frac{\Delta\vartheta}{z_0}$$

and finally

$$\frac{\lambda_s - \lambda}{\lambda} = \frac{2c^2\pi}{a^2\Delta\vartheta} \sum_{n=1}^{\infty} \sum_{m=1}^{\infty} (-1)^n C_{n,m} \frac{n}{\xi_m \beta_{n,m}} \sin\left(\xi_m \frac{a}{c}\right) \sinh\left(\beta_{n,m} \frac{a}{c}\right)$$

Inserting $C_{n,m}$ according to Eq 4*b* yields

$$\frac{\lambda_s - \lambda}{\lambda} = F_1 + \frac{\vartheta_m - \vartheta_u}{\Delta\vartheta} F_2 \tag{9b}$$

Factors F_1 and F_2 (with p according to Eq 5*b*) here signify

$$F_1 = 8p \frac{c^2}{a^2} \sum_{m=1}^{\infty} \frac{\sin(\xi_m) \cdot \sin\left(\xi_m \frac{a}{c}\right)}{[\xi_m^2 + p \cos^2(\xi_m)]} \times \sum_{n=2,4,6\ldots}^{\infty} \frac{\sinh\left(\beta_{n,m} \frac{a}{c}\right)}{\beta_{n,m}^2 \left[\sinh \beta_{n,m} + \frac{p}{\beta_{n,m}} \cosh \beta_{n,m}\right]}, \tag{10b}$$

$$F_2 = 16p \frac{c^2}{a^2} \sum_{m=1}^{\infty} \frac{\sin(\xi_m) \cdot \sin\left(\xi_m \frac{a}{c}\right)}{[\xi_m^2 + p \cos^2(\xi_m)]} \times \sum_{n=1,3,5\ldots}^{\infty} \frac{\sinh\left(\beta_{n,m} \frac{a}{c}\right)}{\beta_{n,m}^2 \left[\sinh \beta_{n,m} + \frac{p}{\beta_{n,m}} \cosh \beta_{n,m}\right]} \tag{11b}$$

and in the limiting case $p \to \infty$ $(a \neq c)$

$$F_1^* = -8 \frac{c^2}{a^2} \sum_{m=1}^{\infty} (-1)^m \frac{\sin\left(\xi_m \frac{a}{c}\right)}{\xi_m^2} \times \sum_{n=2,4,6\ldots}^{\infty} \frac{\sinh\left(\beta_{n,m} \frac{a}{c}\right)}{\beta_{n,m} \cosh \beta_{n,m}} \tag{10b*}$$

$$F_2^* = -16 \frac{c^2}{a^2} \sum_{m=1}^{\infty} (-1)^m \frac{\sin\left(\xi_m \frac{a}{c}\right)}{\xi_m^2} \times \sum_{n=1,3,5\ldots}^{\infty} \frac{\sinh\left(\beta_{n,m} \frac{a}{c}\right)}{\beta_{n,m} \cosh \beta_{n,m}} \tag{11b*}$$

In the last equations the roots of Eq 7 are

$$\xi_m = \frac{(2m - 1)}{2} \pi$$

Analytical Findings

For circular and square plates the model established at the beginning of Section 2 of plate devices with isothermal ends on the test plate including the guard ring region led to the formally equivalent error equation

$$\frac{\lambda_s - \lambda}{\lambda} = F_1 + \frac{\vartheta_m - \vartheta_u}{\Delta\vartheta} F_2 \tag{9}$$

where the error factors F_1 and F_2 are defined by Eq 10 and 11. These depend not only on the dimensionless heat exchange parameter p but also on geometric quantities, which also can be made dimensionless by defining

$$VS = \frac{z_0}{2r_0} \quad \text{or} \quad \frac{z_0}{2a} \tag{10}$$

which is the ratio of the specimen thickness to the plate diameter or to the edge length; and by

$$VB = \frac{r_2 - r_0}{2r_0} \quad \text{or} \quad \frac{c - a}{2a} \tag{11}$$

the ratio of the guard ring width to the plate dimensions as used previously. By introducing these quantities all dimension-related quantities disappear in the error factors F_1 and F_2, and their dependence is limited to the independent variables p, VS, and VB.

We can see that the error is not a constant, as is obtained with two-dimensional approximate calculations, for example, by Woodside [2], but depends not only on F_1 (p, VS, VB) and F_2 (p, VS, VB) (which yield the constant factors F_1 and F_2 for the same installation in a plate device), but also to a decisive extent on the reduced temperature difference $(\theta_m - \theta_u)/\Delta\theta$.

As will be substantiated in later calculations, usually

$$F_1 \ll F_2,$$

in the practical range of use of the three variables p, VS, VB, so that the first error factor can be disregarded

$$\frac{\lambda_s - \lambda}{\lambda} \approx \frac{\vartheta_m - \vartheta_u}{\Delta\vartheta} F_2 \tag{12}$$

With an ambient temperature θ_u which can be adjusted to the mean test temperature $\theta_m = 1/2(\theta_2 + \theta_1)$ the experimenter, therefore, has the possibility of making the boundary error as small as the uncertainties of individual temperature measurements will permit. On the other hand, if the ambient temperature is not considered, the test-constant error factor F_2 can be multiplied under certain conditions by an unanticipatedly large positive or negative term representing reduced temperature differences.

It is worth noting that according to Eq 12 the error can be minimized by reasonable adjustment of the ambient temperature θ_u, even without a quantitative knowledge of the factor F_2.

Experimental Checks

In view of the derived results, our own double plate apparatus (where $r_0 =$ 61.0 mm and $r_2 - r_0 = 41.0$ mm) as shown in Fig. 2 was equipped with an additional copper shield around the specimen and soldered to a copper coil which permitted us to regulate the shielding by means of thermostatically controlled circulating water so that it was at a measurable and adjustable temperature (the "ambient temperature" θ_u). Between the specimen, including hot and cold plates, and the "environment" was inserted thermal insulating material of loosely packed cotton wads, for the purpose of avoiding convection and preventing radiation.

The approximate Eq 12, when solved for λ_s (the measured thermal conductivity)

$$\lambda_s = \lambda + \lambda F_2(p, VS, VB) \cdot \frac{\vartheta_m - \vartheta_u}{\Delta\vartheta} \tag{13}$$

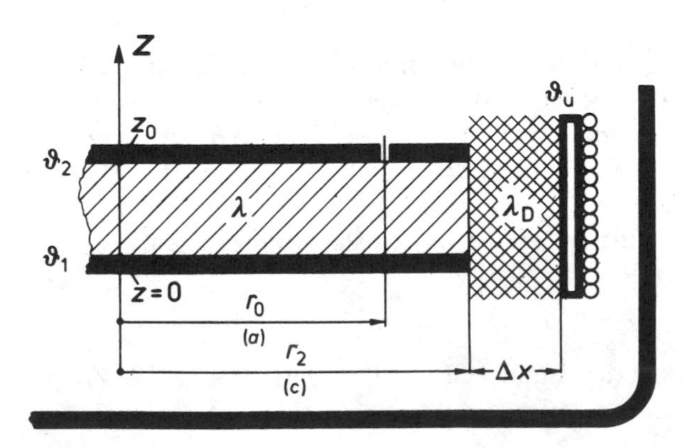

FIG. 2—*Modified plate apparatus with isothermal copper shield which makes it possible to adjust the ambient temperature* θ_u.

yields a linear dependence of the measured values λ_s on the reduced temperature difference $(\theta_m - \theta_u)/\Delta\theta$, if during the variation of the latter the mean temperature θ_m and, therefore, $\lambda(\theta_m)$ are kept constant.

This analytical relationship was checked experimentally by controlled variations of the ambient temperature θ_u using the supplementary device described above, whereby θ_m was kept constant. In order to make the expected effect as great as possible, the highest possible test temperature which was still permissible with respect to materials with $\theta_m = 60°C$ was selected. The results shown in Fig. 3 confirm the linear dependence of the measured thermal conductivities λ_s on the reduced temperature difference. The true thermal conductivity of specimen λ is obtained at the point of intersection of the plotted approximation lines with the ordinate at $(\theta_m - \theta_u)/\Delta\theta = 0$.

In addition it is possible to determine the slope of the lines which according to Eq 13 are supposed to equal λF_2; after division by the now known λ, we obtain the experimental F_2. In a comparison with previously calculated $F_2(p, VS, VB)$ values for the established geometric quantities and for p values of 0.1 to ∞ we find agreement of the calculated and experimental factor F_2 at p between the numerical values 2 and 3. From p it is again possible to calculate the thermal conductivity λ_D of the thermal insulating material from the equation

$$p = \frac{\alpha r_2}{\lambda} \approx \frac{\lambda_D/\lambda}{\ln\left[1 + \dfrac{\Delta x}{r_2}\right]}$$

which conductivity here yields the plausible quantity for the thermal conductivity of air. (In this special example $F_2 \approx 0.02$ and $F_1 \approx 0.0001$.)

In a similar way measurements were evaluated in which there was no ther-

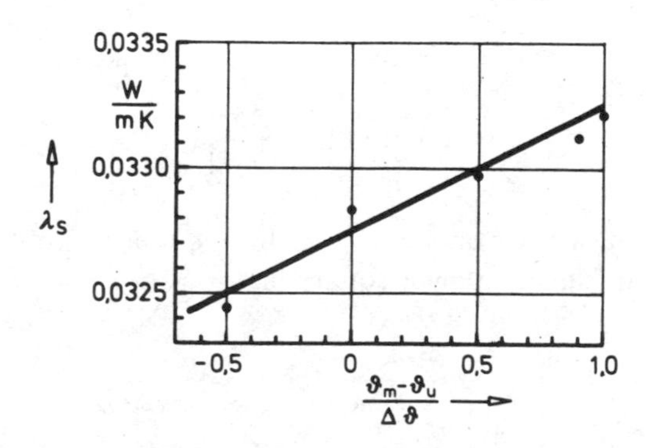

FIG. 3—*Measured thermal conductivity values λ_s of extruded foam over the reduced temperature difference. (θ_m = mean, $\Delta\theta$ = difference between hot and cold plate temperatures; θ_u = ambient temperature). Measured values approximated by straight line.*

mal insulation between the edge of the specimen and the copper shielding. The wide air gap led us to expect convection. Indeed the experiments yielded α values from 9 to 10 W m^{-2} K^{-1}, which correspond more or less to natural convection along vertical walls.

On the whole these experimental results show that the models lead to a good approximation of real conditions.

Numerical Calculations

The error factors F_1 and F_2 in Eq 9 are a function of the variables VS and VB given by the apparatus which have an unchanged effect for the same apparatus and the same specimen. What remains is the dependence on the heat exchange parameter p according to Eqs 5a and 5b, the quantity of which will first be estimated.

The Heat Exchange Parameter

$p = 0$ means complete thermal insulation on the outer edge of the guard ring: the temperature field in the specimen and in the guard ring remains understood, and the overall error disappears ($F_1 = F_2 = 0$). If no insulating material is used along the edge and if air is left, then convection takes place, as the experiments described above show; the convection, depending on the device size, leads to relatively large p-values (approximately 20, 30, 50). In the calculations for F_1 and F_2 it is apparent that with supposed convection (with practically unknown α) p can be approximated by ∞ without a great calculation error. That leads to somewhat higher but more reliable F_1 and F_2 values.

If an insulating material with a thermal conductivity λ_D is used α can be replaced in good approximation by

$$\alpha \approx \frac{\lambda_D}{r_2 \ln\left[1 + \dfrac{\Delta x}{r_2}\right]}$$

for the case involving a circle or $\alpha \approx \lambda_D/\Delta x$ for the square problem. Then the following is true for the circular or square plate

$$p \approx \frac{\lambda_D/\lambda}{\ln\left[1 + \dfrac{\Delta x}{r_2}\right]} \text{ or } p \approx \frac{\lambda_D/\lambda}{\Delta x/c} \tag{14}$$

For the latter case the relationships are shown in a diagram in Fig. 4.

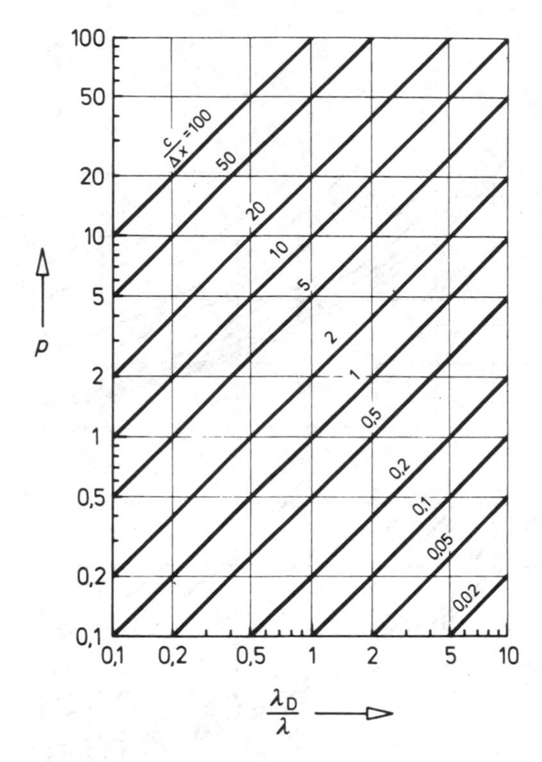

FIG. 4—*Heat exchange parameter* p *for square plates when a thermal insulating material is used around the test setup.* λ = *thermal conductivity of the test specimen,* λ_D = *thermal conductivity of the insulating material,* c = *half the test plate edge length (guard ring outside dimension),* Δx = *distance between the outer edge of the guard ring and the "environment."*

Error Factor F_2

Figure 5 gives a rough overview of the orders of magnitude which occur. F_2 is plotted in the figure on a logarithmic scale for the reduced guard ring widths VB = 0.25, 0.375, 0.5, over the reduced specimen widths VS; the parameter p is varied from = 0.1 to ∞.

The calculated values apply to the square plate apparatus. This representation was selected in consideration of the most widespread working method involving the square device type. The corresponding values for the circular plate devices differ from these only slightly if the same VS, VB and p values are valid. In the drawings the differences are generally covered over by the thickness of the curve line. In general the numerical values in the case of the circle are somewhat greater than those of the square plate. As was already mentioned, in the range of validity of the approximately $F_1 \ll F_2$, which will be discussed below, F_2 is representative of the boundary error which is not

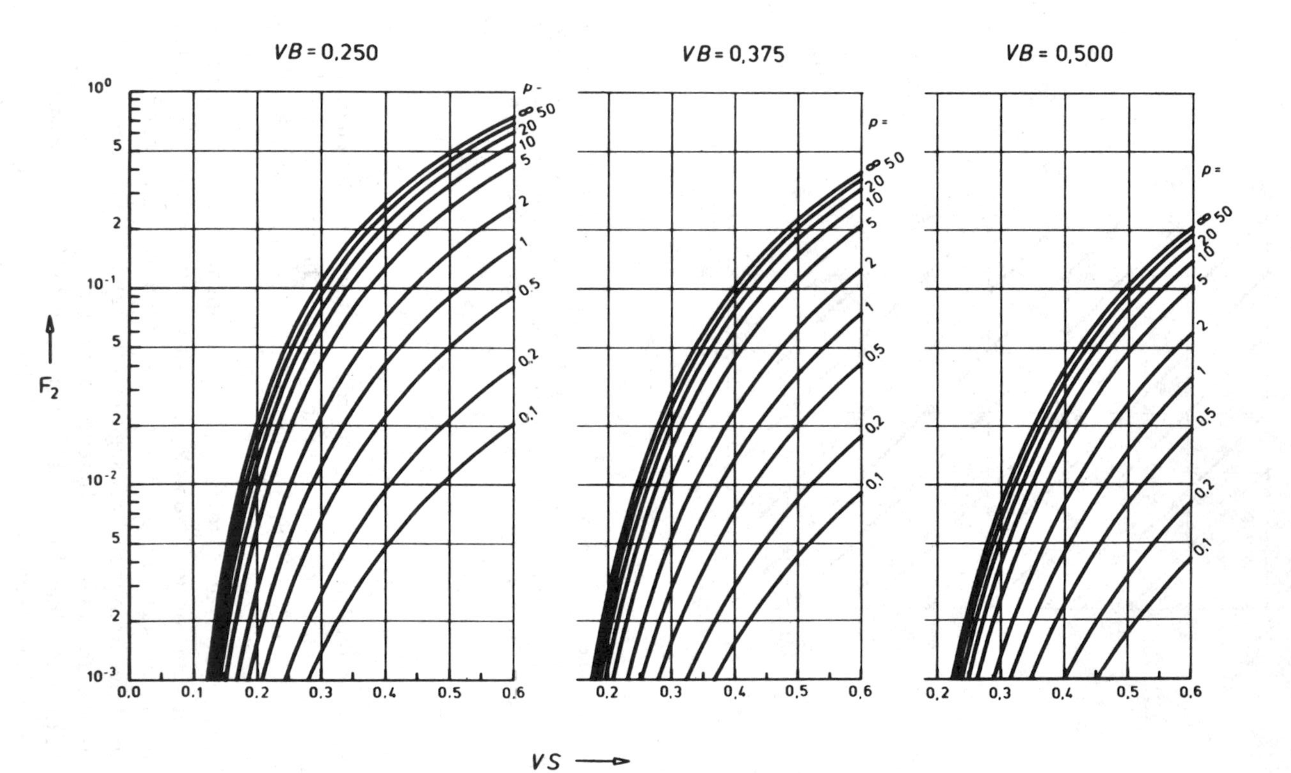

FIG. 5—*Error factor* F_2 *(*VS, VB, p*) over the reduced plate height* VS *(plate height over edge length* 2a*) for three reduced guard ring widths* VB *(guard ring width over edge length) with parameter values* p.

only a function of the geometric relationships, since F_2 must be also multiplied by the reduced temperature difference.

As was to be expected, the smaller the guard ring width and the greater the specimen thickness, the greater the error factor; furthermore, the error increases with increasing p values. $p \rightarrow \infty$, inserted as the maximum limit, leads to a reliable estimate.

Figure 6 is intended to provide information about the VS value (specimen height over plate length) which may not be exceeded if at a given VB (guard ring width over plate length) a desired F_2 value is to be preserved. In this case the heat exchange parameter $p \rightarrow \infty$ is assumed. Example: at $VB = 0.25$, F_2 should not become greater than 5%; this is guaranteed as long as $VS <$ 0.245. For the apparatus conforming to the standard DIN 52 612 [3] with an edge length of 500 mm and a guard ring width of 125 mm (that is, $VB =$ 0.25), this would yield a maximum specimen height of 0.245×500 mm = 122.5 mm.

Table 1 is to be treated in a similar manner. In this case the fixed error factor $F_2 = 1\%$ is given, and the respective VS is tabulated for customary VB values and the possible p variations.

Measurements Without a Guard Ring

In spite of the well-known susceptibility of measurements without a guard ring to disturbances, we find in current literature again and again reports on such studies. Misjudging the realities, these studies mistakenly characterize

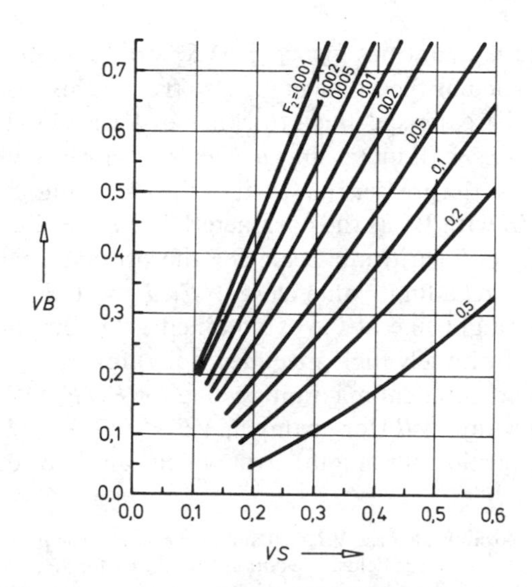

FIG. 6—*Curves of the constant error factor* F_2 *for the heat exchange parameter* $p \rightarrow \infty$ *in VB* (ordinate) — VS (abscissa) *diagram.*

TABLE 1—*Maximum* VS *values (specimen height over edge length) with the quadratic plate for error factor* $F_2 = 1\%$; VB: *guard ring width over edge length;* p: *heat exchange parameter according to Eq* 5b.

	$VB =$	0.20	0.25	0.30	0.35	0.40	0.45	0.50	0.55	0.60	0.65	0.70	0.75
$p = \infty$	$VS =$	0.15	0.17	0.20	0.23	0.26	0.28	0.31	0.34	0.36	0.39	0.41	0.44
50.0		0.15	0.18	0.21	0.24	0.27	0.29	0.32	0.35	0.37	0.40	0.42	0.45
20.0		0.16	0.19	0.22	0.25	0.28	0.30	0.33	0.36	0.38	0.41	0.44	0.46
10.0		0.17	0.20	0.23	0.26	0.29	0.32	0.35	0.37	0.40	0.43	0.46	0.48
5.0		0.19	0.22	0.25	0.28	0.31	0.34	0.37	0.40	0.43	0.46	0.48	0.51
2.0		0.22	0.25	0.29	0.32	0.35	0.38	0.42	0.45	0.48	0.51	0.54	0.57
1.0		0.25	0.29	0.32	0.36	0.39	0.43	0.46	0.50	0.53	0.56	0.59	0.63
0.5		0.29	0.33	0.37	0.41	0.45	0.48	0.52	0.56	0.59	0.63	0.67	0.70
0.2		0.36	0.41	0.45	0.50	0.54	0.59	0.63	0.67	0.71	0.75	0.79	0.83
0.1		0.43	0.49	0.54	0.59	0.64	0.69	0.74	0.78	0.83	0.88	0.92	0.97

materials with low thermal conductivity, in particular, as favorable, although it is precisely these which lead to a high p value (see Eq 5).

Equations 10 and 11 also allow us to include the case in which there is no guard ring ($VB = 0$ or $r_2 = r_0$ or $c = a$). However, Eqs 10* and 11* (for $p \to \infty$) must be excluded, since they lead to divergent series.[4] The results of calculations for the error factor F_2 are shown in Fig. 7. The quantitative order of magnitude of F_2 urgently underscores the susceptibility to error which can be expected in measurements taken without a guard ring.

Error Factor F_1

The totality of the quantities for F_1 (p, VS, VB) is qualitatively similar to that for F_2; in particular the following is also true in this case: a small guard ring width VB and greater specimen thicknesses VS lead to higher F_1 values; likewise, F_1 increases with increasing p. However, F_1 is always smaller than F_2. The greater p is, the greater the fraction F_1/F_2. In Fig. 8 F_1/F_2 is plotted for $p \to \infty$ over VB with VS as the parameter. The F_1/F_2 lines are practically straight lines with a logarithmic scale for the ordinate, whereas those for $VB = 0$ tend towards a limit value close to $F_1/F_2 = 0.5$.

A horizontal straight line at 0.5 is obtained as a computational limiting case for $VS \to \infty$ (infinitely high specimen body) for F_1/F_2. This value is at the same time the theoretical maximum value for F_1/F_2. With insufficiently small guard ring widths VB (for example, $VB < 0.25$) F_1/F_2 is still so large that F_1 leads to a portion of the total error which cannot be disregarded. This

[4]Whereas the error formulas in case $VB = 0$ and at the same time $p \to \infty$ fail, the series in Eq 3 converges for the temperature fields. We obtain in that case for $\theta(r, z)$ or $\theta(x, y, z)$ or both a temperature distribution which goes over into the constant temperature θ_u on the outer edge of the guard ring.

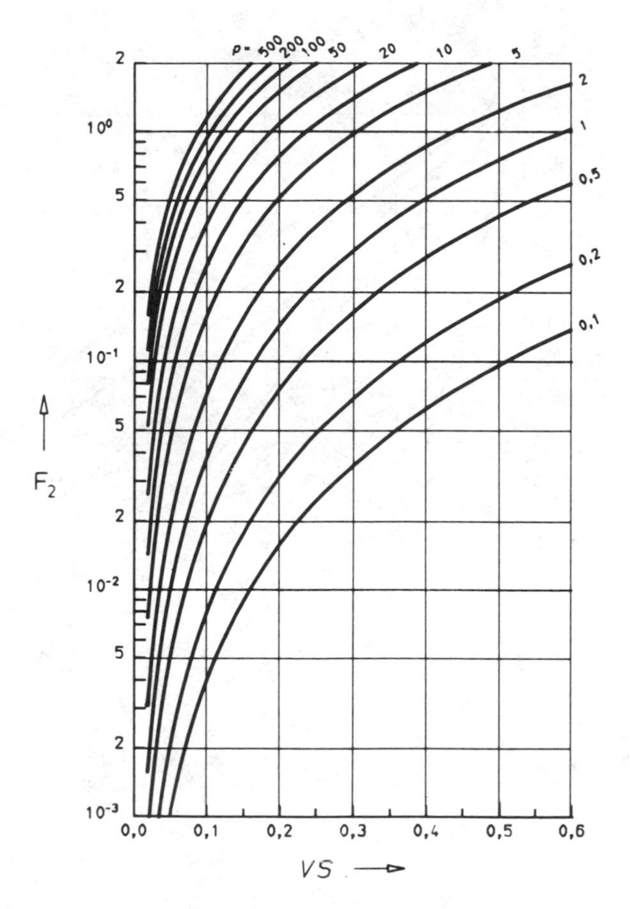

FIG. 7—*Error factor* F_2 *over the reduced plate height* VS *for plate apparatus without a guard ring (*VB = 0*). Heat exchange parameter* p. *(For* p → ∞, F_2 *can become any size.)*

is a fact which must be seriously considered in evaluating the error for measurements without a guard ring.

Above $VB = 0.25$, the numerical value of F_1/F_2 is always so small, if VS is reasonably selected that there is sufficient support for disregarding the constant error contribution F_1 (VS, VB, p).

The broken lines in Fig. 8 are curves for the constant error factor F_2. This diagram shows that when there is a practical effort made to sufficiently minimize the influential error factor F_2 (for example, 1%), F_1/F_2 is only a few thousandths, and that on the other hand large F_2 values (for example, 20%), which already make the measurement worthless, involve in addition high F_1 values.

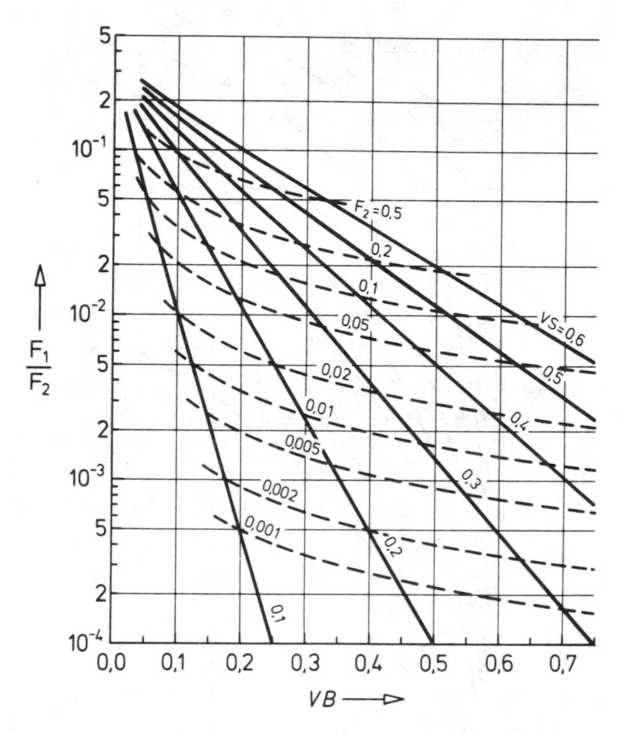

FIG. 8—*Quotient* F_1/F_2 *of the error factors over the reduced guard ring width for* $p \rightarrow \infty$*; parameter (*solid lines*) reduced specimen height* VS. *The broken curves are the lines of the constant error factor* F_2.

Acknowledgments

Mr. H. W. Krupke did most of the programming for the numerical calculations; Mr. R. Jugel took care of the experimental studies. I am very grateful to both colleagues.

References

[1] Poensgen, R., *Z. Ver. dt. Ing.*, Vol. 56, 1912, p. 1653, and reports on research work in the area of engineering, No. 130, 1912.

[2] Woodside, W., *Symposium on Thermal Conductivity Measurements and Applications of Thermal Insulations*, *STP 217*, American Society for Testing and Materials, Philadelphia, 1957, p. 49.

[3] German Standard DIN 52 612, Part 1, Determination of Thermal Conductivity with the Plate Device.

Kazuo Eguchi[1]

Error Analysis of Measuring Thermal Conductivity by Circular and One Side Heat Flow Type Guarded Hot Plate Apparatus

REFERENCE: Eguchi, K., **"Error Analysis of Measuring Thermal Conductivity by Circular and One Side Heat Flow Type Guarded Hot Plate Apparatus,"** *Guarded Hot Plate and Heat Flow Meter Methodology, ASTM STP 879*, C. J. Shirtliffe and R. P. Tye, Eds., American Society for Testing and Materials, Philadelphia, 1985, pp. 49–68.

ABSTRACT: This paper deals with the theoretical analysis on total errors which occur upon measuring the thermal conductivity of a circular specimen using a one side heat flow type guarded hot plate.

Introducing three input parameters, that is, (*a*) temperature of ambient air, (*b*) that of guard ring, and (*c*) that of guard plate, which are normalized by letting the temperature of a hot plate be a unit and that of the cooling plate be zero, the theoretical solution of the total errors was derived. The errors were classified into the ones caused by the three parameters, and they were also classified according to the heat flow paths.

Heat exchange at the edges of the specimen and insulation between hot plate and guard plate was treated as the surface heat transfer including edge insulation.

The total errors were expressed with the following seven normalized parameters, that is, (*a*) ratio of the radius of the hot plate to that of the guard ring, (*b*) ratio of the thickness of the specimen to the radius of the guard ring, (*c*) ratio of the thickness of the insulation between the hot plate and the guard plate to the radius of the guard ring, (*d*) ratio of the thermal conductivity of the insulation to that of the specimen, (*e*) ratio of the equivalent edge insulation thickness to the radius of the guard ring, (*f*) ratio of the direct heat flow across the gap to the heat flow out of the hot plate into the specimen which is obtained assuming one-directional heat flow, and (*g*) ratio of the direct heat flow through the thermal bridges between the hot plate and the guard plate to the heat flow out of the hot plate into the insulation which is obtained assuming one-directional heat flow.

In this paper, each of (*f*) and (*g*) is considered to be determined by the analytical solution using experimental data.

KEY WORDS: guarded hot plate, heat flow meter, oneside heat flow, circular specimen, theoretical analysis, total errors, reciprocity theorem

[1]Principal research director, Building Research Institute, Ministry of Construction, Tsukuba, Japan.

Nomenclature

HP Hot plate
CP Cooling plate
GR Guard ring
GP Guard plate
T_H Temperature of HP (K)
T_c Temperature of CP (K)
T_g Temperature of GR (K)
T_p Temperature of GP (K)
T_a Temperature of ambient air (K)
g Normalized temperature of GR (dimensionless), as defined in Eq 4
p Normalized temperature of GP (dimensionless), as defined in Eq 5
a Normalized temperature of ambient air (dimensionless), as defined in Eq 6
r, Z Cylindrical coordinates
r_0 Radius of GR (m)
r_s Radius of HP (m)
d_0 Radial thickness of edge insulation (m)
d_1 Equivalent-edge-insulation thickness (m), as defined in Eq 2
ℓ_1 Thickness of the specimen (m)
ℓ_2 Thickness of the insulation between HP and GP (m)
h Heat transfer coefficient ($W/m^2 K$)
α Convection heat transfer coefficient (W/m^2K)
ϕ_g Heat transfer coefficient for direct heat transfer across the gap (W/K)
ϕ_p Heat transfer coefficient for direct heat transfer through the thermal bridge between HP and GP (W/K)
λ_1 Thermal conductivity of specimen ($W/m K$)
λ_2 Thermal conductivity of insulation between HP and GP (W/m K)
C_1 Equivalent-edge-insulation parameter (dimensionless), as defined in Eq 3
C_2 Abbreviation according to Eq 41
λ_3 Thermal conductivity of edge insulation ($W/m K$)
λ_s Measured thermal conductivity of specimen ($W/m K$)
η Dimensionless parameter as defined Eq 40
R_s Dimensionless parameter as defined Eq 17

L_1 Dimensionless parameter as defined Eq 18
L_2 Dimensionless parameter as defined Eq 39
$\theta_1,\theta_2,\theta_3,\theta_4,\theta_5,\theta_6,\theta_7,\theta_8,\theta_9$ Temperature distribution under the condition shown in Fig. 3*a* and *b* (dimensionless
$Q_1,Q_2,Q_3,Q_4,Q_5,Q_6,Q_7,Q_8,Q_9$ Heat flow rate at HP surfaces under the condition shown in Fig. 3*a* and *b* (W/K)
q_g, q_p Heat flow rate under the condition shown in Fig. 3*c* (W/K)
q_1 Heat flow rate as defined Eq 21 (W/K)
q_2 Heat flow rate as defined Eq 30 (W/K)
$\bar{Q}$ Rate of the total heat input to HP under the conditions shown in Fig. 2*a, b*, and *c* (W/K)
Q_0 Rate of the one-directional heat flow across the cylindrical specimen of radius r_s, under the $1K$ difference in temperatures between HP and CP (W/K)
X_1 Abbreviation according to Eq 19
Y_1 Abbreviation according to Eq 20
Z_1 Abbreviation according to Eq 23
X_2 Abbreviation according to Eq 42
Y_2 Abbreviation according to Eq 43
Z_2 Abbreviation according to Eq 44
W Abbreviation according to Eq 45
A_0 Abbreviation according to Eq 58
A_1 Abbreviation according to Eq 59
A_2 Abbreviation according to Eq 60
G_1 Abbreviation according to Eq 62
G_2 Abbreviation according to Eq 63
G_3 Abbreviation according to Eq 64
P_2 Abbreviation according to Eq 66
P_3 Abbreviation according to Eq 67
J_0, J_1 Bessel function of the first kind
m_k k^{th} root of Eq 16
n_k k^{th} root of Eq 38
$\bar{\epsilon}_T$ Total error due to the difference in temperature between HP and GR, and between HP and GP, and due to edge heat losses
$\bar{\epsilon}_g$ Total error due to the difference in temperature between HP and GR
ϵ_{g1} Error due to the difference in temperature between HP and GR which varies depending on the heat transfer in the specimen
ϵ_{g2} Error due to the difference in temperature between HP and GR which varies depending on

the heat transfer in the insulation material between GP and HP

ϵ_{g3} Error due to the difference in temperature between HP and GR which varies depending on the direct heat transfer across the gap

$\overline{\epsilon}_p$ Total error due to the difference in temperature between HP and GP

ϵ_{p2} Error due to the difference in temperature between HP and GP which varies depending on the heat transfer in the insulation material between GP and HP

ϵ_{p3} Error due to the difference in temperature between HP and GP which varies depending on the direct heat transfer across the thermal bridges between GP and HP (structural connections and lead wires, etc.)

$\overline{\epsilon}_a$ Total error due to edge heat losses

ϵ_{a1} Error due to edge heat losses of the specimen

ϵ_{a2} Error due to edge heat losses of the insulation material between GP and HP

Measuring methods for thermal properties of building materials and components have been studied at the Building Research Institute (BRI) since 1979. One of the main purposes of the study is to develop methods and apparatuses for measuring thermal properties accurately, reproducibly, and rapidly. As a part of the project, the circular and one side heat flow type guarded hot plate (GHP) has been developed. The apparatus is called a "single specimen apparatus," and a diagram of the apparatus is shown in Fig. 1. In general, thermal conductivity is measured with a GHP apparatus assuming one-directional heat flow. One-directional heat flow, however, is hardly realized be-

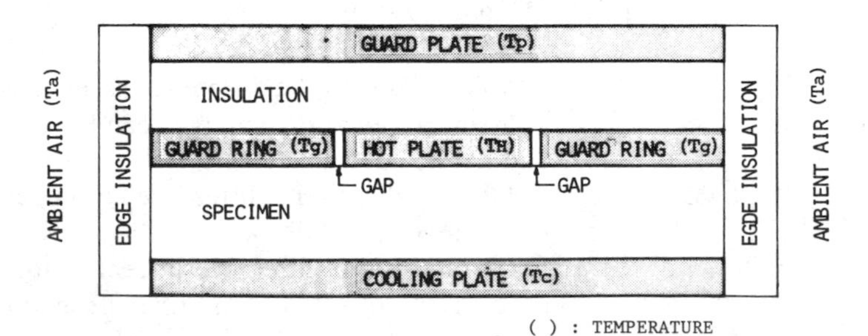

FIG. 1—*Diagram of single specimen guarded hot plate apparatus.*

cause of heat exchange with ambient air at the edge surface, the difference in temperature between the hot plate (HP) and the guard ring (GR) and the difference in temperature between HP and guard plate (GP).

Lack of one-dimensional heat flow causes systematic errors. The deviation depends on the dimension of the apparatus, thermal conductivity of insulation between HP and GP, mechanical detail of the apparatus to cause direct error heat flow, thickness and thermal conductivity of specimen and edge insulation, temperature differences caused by controlling errors, and ambient condition. Because of the large number of parameters involved, theoretical analyses are needed to estimate systematic errors. Heat losses due to direct heat flow across the gap between the HP and GR and the thermal bridges between HP and GP caused by the mechanical structure are hardly estimated by a theoretical analysis. Experiments are still necessary to evaluate these effects.

This paper deals with the theoretical analysis which is one of the two processes to determine the parameters to calculate the error and to correct the measured conductivity. The experiments are now being conducted at BRI and will be reported in the near future.

Theoretical Analysis

Many researchers have already discussed theoretical assessments on heat conduction of square and two specimen type GHP [*1-6*]. However, no one has studied that of circular and one side heat flow type GHP.

Assuming that the temperature of the specimen edge is equal to that of the cooling plate, Somers and Cyphers [*1*] have analyzed the circular and two specimens type GHP.

The author and Miyake [*7*] have analyzed the circular and one side heat flow type GHP, using the equivalent edge temperature which was introduced by Woodside [*3*] for the square GHP.

In this paper by improving boundary conditions as realisticly as possible, the analysis in the previous report [*7*] is modified as follows:

1. The heat flow in the insulation material which is due to the difference in temperature between the HP and GP is handled as the three-dimensional heat flow expressed in cylindrical coordinates, whereas we assumed one-dimensional heat flow in the previous report [*7*].
2. The heat exchange at the edge of the specimen and insulation between HP and GP are analyzed as the surface heat transfer caused by the difference in temperature between the ambient and the edge surface, whereas we assumed a uniform temperature (equivalent-edge-temperature) [*3*] on the edge surface of the specimen in the previous report [*7*].

In order to make a mathematical model, following assumption have been used.

(*a*) The specimen is a disk whose radius is, r_0, thermal conductivity, λ_1, and thickness, ℓ_1. The upper surface is in contact with HP and GR, and the lower surface is in contact with cooling plate (CP).

(*b*) Insulation between GP and HP is a disk whose radius is r_0, thermal conductivity, λ_2, and thickness, ℓ_2, and the upper surface is in contact with GP, and the lower surface is in contact with HP and GR.

(*c*) The temperatures of isothermal surfaces at HP, GR, GP, and CP are maintained at T_H, T_g, T_p, and T_c, respectively.

(*d*) The width of gap separating the HP and GR is negligibly small.

(*e*) The heat flow rate across the cylindrical boundary edge surface of both specimen and the insulation between HP and GP is proportional to the difference in temperature between the edge and constant ambient air T_a. Heat transfer coefficient h is constant irrespective of temperature.

(*f*) The error heat flow directly across the gap separating the HP and GR is proportional to the difference in temperature between HP and GR. The heat transfer coefficient is denoted at ϕ_g.

(*g*) The error heat flow through the thermal bridges between HP and GP is proportional to the difference in temperature between HP and GP and is independent of temperature distribution in the insulation between HP and GP. The heat transfer coefficient is denoted as ϕ_p.

(*h*) Thermal conductivity, λ_1, of the specimen and, λ_2, of the insulation between HP and GP are both constant irrespective of temperature. If the edge insulation of radial thickness, d_0, and thermal conductivity, λ_3, are in contact with the outer edges of both the specimen and insulation between HP and GP, and if the convection heat-transfer coefficient at the exposed edge surface is α, magnitude of h (assumption (*e*)) is calculated from the following equation assuming radial heat flow in the edge insulation

$$h = 1/r_0\left\{\frac{1}{\alpha(r_0 + d_0)} + \frac{1}{\lambda_3}\,\ell\text{n}\left(\frac{r_0 + d_0}{r_0}\right)\right\} \tag{1}$$

The equivalent edge insulation thickness d_1 is defined as

$$d_1 = \lambda_1/h \tag{2}$$

And the equivalent edge insulation parameter C_1 is defined as

$$C_1 = r_0 h/\lambda_1 = r_0/d_1 \tag{3}$$

By applying normalized temperatures (based on $T_H = 1$, $T_c = 0$), the heat flow in the specimen and in the insulation between HP and GP, under the condition shown in Fig. 1 may be simplified to that shown in Fig. 2*a* and *b*, where normalized temperatures are

$$g = (T_g - T_c)/(T_H - T_c) \tag{4}$$

$$p = (T_p - T_c)/(T_H - T_c) \tag{5}$$

$$a = (T_a - T_c)/(T_H - T_c) \tag{6}$$

Figure 2*c* shows the error heat flows that is considered to be independent of temperature distribution under the condition shown in Fig. 2*a* and *b*.

The mathematical model shown in Fig. 2*a* are described by the following set of equations

$$\frac{\partial^2 \theta_1}{\partial r^2} + \frac{1}{r}\frac{\partial \theta_1}{\partial r} + \frac{\partial^2 \theta_1}{\partial Z^2} = 0 \qquad \begin{matrix} 0 < Z < \ell_1 \\ 0 \leqq r < r_0 \end{matrix} \tag{7a}$$

$$\theta_1(r, \ell_1) = 1 \qquad 0 \leqq r < r_s \tag{7b}$$

$$\theta_1(r, \ell_1) = g \qquad r_s < r < r_0 \tag{7c}$$

$$\theta_1(r, 0) = 0 \qquad 0 \leqq r < r_0 \tag{7d}$$

$$-\lambda_1 \left(\frac{\partial \theta_1}{\partial \theta} \right)_{r=r_0} = h\{\theta_1(r_0, Z) - a\} \qquad 0 < Z < \ell_1 \tag{7e}$$

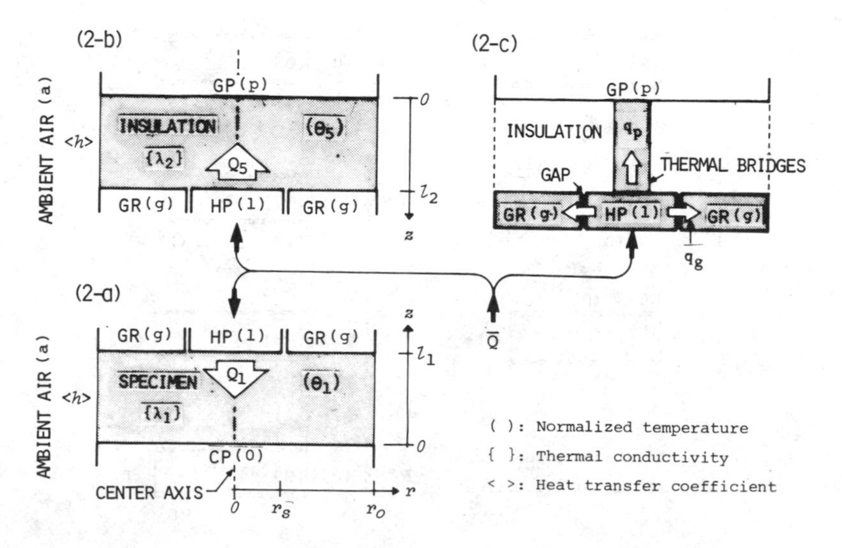

FIG. 2—*Concept of heat conduction in the guarded hot plate apparatus with normalized temperature.*

$$Q_1 = 2\pi\lambda_1 \int_0^{r_s} r\left(\frac{\partial \theta_1}{\partial Z}\right)_{Z=\ell_1} dr \tag{7f}$$

The solutions of heat flow Q_1 may be simplified as follows (refer to Fig. 3*a*)

$$Q_1 = Q_2 - gQ_3 - a \cdot Q_4 \tag{8}$$

where Q_2, Q_3, and Q_4 are the solutions of the following three sets of equations, respectively

$$\frac{\partial^2\theta_2}{\partial r^2} + \frac{1}{r}\cdot\frac{\partial\theta_2}{\partial r} + \frac{\partial^2\theta_2}{\partial Z^2} = 0 \qquad \begin{matrix} 0 < Z < \ell_1 \\ 0 \leqq r < r_0 \end{matrix} \tag{9a}$$

$$\theta_2(r, \ell_1) = 1 \qquad 0 \leqq r < r_s \tag{9b}$$

$$\theta_2(r, \ell_1) = 0 \qquad r_s < r < r_0 \tag{9c}$$

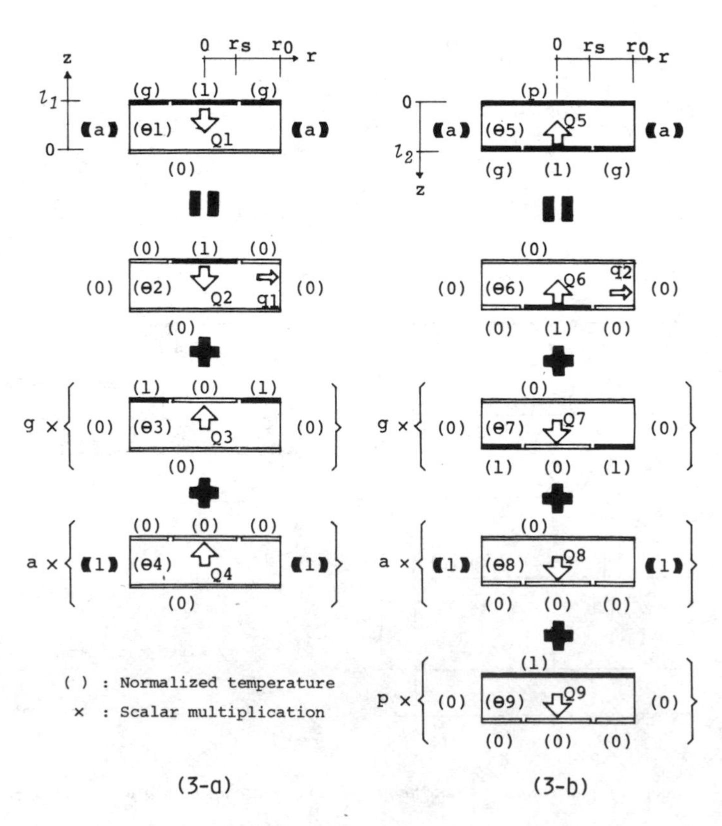

FIG. 3—*Simplification of Fig.* 2a *and* b *for the error analysis.*

$$\theta_2(r, 0) = 0 \qquad 0 \leqq r < r_0 \tag{9d}$$

$$-\lambda_1 \left(\frac{\partial \theta_2}{\partial r} \right)_{r=r_0} = h \cdot \theta_2(r_0, Z) \qquad 0 < Z < \ell_1 \tag{9e}$$

$$Q_2 = 2\pi\lambda_1 \int_0^{r_s} r \cdot \left(\frac{\partial \theta_2}{\partial Z} \right)_{Z=\ell_1} dr \tag{9f}$$

$$\frac{\partial^2 \theta_3}{\partial r} + \frac{1}{r} \cdot \frac{\partial \theta_3}{\partial r} + \frac{\partial^2 \theta_3}{\partial Z^2} = 0 \qquad \begin{matrix} 0 < Z < \ell_1 \\ 0 \leqq r < r_s \end{matrix} \tag{10a}$$

$$\theta_3(r, \ell_1) = 0 \qquad 0 \leqq r < r_s \tag{10b}$$

$$\theta_3(r, \ell_1) = 1 \qquad r_s < r < r_0 \tag{10c}$$

$$\theta_3(r, 0) = 0 \qquad 0 \leqq r < r_0 \tag{10d}$$

$$-\lambda_1 \left(\frac{\partial \theta_3}{\partial r} \right)_{r=r_0} = h \cdot \theta_3(r_0, Z) \qquad 0 < Z < \ell_1 \tag{10e}$$

$$-Q_3 = 2\pi\lambda_1 \int_0^{r_s} r \cdot \left(\frac{\partial \theta_3}{\partial Z} \right)_{Z=\ell_1} dr \tag{10f}$$

$$\frac{\partial^2 \theta_4}{\partial r^2} + \frac{1}{r} \cdot \frac{\partial \theta_4}{\partial r} + \frac{\partial^2 \theta_4}{\partial Z^2} = 0 \qquad \begin{matrix} 0 < Z < \ell_1 \\ 0 \leqq r < r_0 \end{matrix} \tag{11a}$$

$$\theta_4(r, \ell_1) = 0 \qquad 0 \leqq r < r_0 \tag{11b}$$

$$\theta_4(r, 0) = 0 \qquad 0 \leqq r < r_0 \tag{11c}$$

$$-\lambda_1 \left(\frac{\partial \theta_4}{\partial r} \right)_{r=r_0} = h \{ \theta_4(r_0, Z) - 1 \} \qquad 0 < Z < \ell_1 \tag{11d}$$

$$-Q_4 = 2\pi\lambda_1 \int_0^{r_s} r \cdot \left(\frac{\partial \theta_4}{\partial Z} \right)_{Z=\ell_1} dr \tag{11e}$$

The solutions of θ_2, Q_2, θ_3, and Q_3 are easily obtained as follows [8]

$$\theta_2(r, Z) = 2\frac{r_s}{r_0}\sum_{k=1}^{\infty}\frac{m_k}{m_k^2 + C_1^2}\frac{J_1(m_kR_s)}{[J_0(m_k)]^2}\frac{\sinh(m_kZ/r_0)J_0(m_kr/r_0)}{\sinh(m_kL_1}\tag{12}$$

$$Q_2 = 4\pi\lambda_1 r_s R_s X_1\tag{13}$$

$$\theta_3(r, Z) = 2\frac{1}{r_0}\sum_{k=1}^{\infty}\frac{m_k}{m_k^2 + C_1^2} \cdot \frac{\{r_0J_1(m_k) - r_sJ_1(m_kR_s)\}}{[J_0(m_k)]^2\sinh(m_kL_1)}\sinh(m_kZ/r_0)J_0(m_kr/r_0)\tag{14}$$

$$Q_3 = 4\pi\lambda_1 r_s(R_sX_1 - Y_1)\tag{15}$$

where $\{m_k\}$ are the roots of

$$m_kJ_1(m_k) - C_1J_0(m_k) = 0\tag{16}$$

where J_0 and J_1 are the Bessel functions of the first kind, and

$$R_s = r_s/r_0\tag{17}$$

$$L_1 = \ell_1/r_0\tag{18}$$

$$X_1 = \sum_{k=1}^{\infty}\frac{m_k}{m_k^2 + C_1^2}\left\{\frac{J_1(m_kR_s)}{J_0(m_k)}\right\}^2\frac{1}{\tanh(m_kL_1)}\tag{19}$$

$$Y_1 = \sum_{k=1}^{\infty}\frac{m_k}{m_k^2 + C_1^2}\frac{J_1(m_k)J_1(m_kR_s)}{[J_0(m_k)]^2}\frac{1}{\tanh(m_kL_1)}\tag{20}$$

Since only the heat flow, Q_4, is needed for error analysis and the magnitude of heat flow, Q_4, is equal to that of heat flow at edge surface of specimen, q_1, under the condition described by Eqs 9*a* through *e* (refer to Fig. 3*a*), applying the reciprocity theorem [9], the heat flow, Q_4, is easily solved using $\theta_2(r, Z)$ in Eq 12, without the solution of $\theta_4(r, Z)$.

Thus,

$$Q_4 = q_1 = -2\pi\lambda_1 r_0\int_0^{\ell_1}\left(\frac{\partial\theta_2}{\partial r}\right)_{r=r_0}dZ\tag{21}$$

Substituting Eq 12 into Eq 21

$$Q_4 = 4\pi\lambda_1 r_s Z_1 \tag{22}$$

where

$$Z_1 = \sum_{k=1}^{\infty} \frac{m_k}{m_k^2 + C_1^2} \frac{J_1(m_k)J_1(m_k R_s)}{[J_0(m_k)]^2} \tanh(m_k L_1/2) \tag{23}$$

Then, substituting Eqs 13, 15, and 21 into Eq 8, the solution of Q_1 is

$$Q_1 = 4\pi\lambda_1 r_s\{R_s X_1(1-g) + gY_1 - aZ_1\} \tag{24}$$

In the same way, a mathematical model shown in Fig. 2*b* are described by the following set of equations

$$\frac{\partial^2\theta_5}{\partial r^2} + \frac{1}{r}\frac{\partial\theta_5}{\partial r} + \frac{\partial^2\theta_5}{\partial Z^2} = 0 \qquad \begin{matrix} 0 < Z < \ell_2 \\ 0 \leqq r < r_0 \end{matrix} \tag{25a}$$

$$\theta_5(r, \ell_2) = 1 \qquad 0 \leqq r < r_s \tag{25b}$$

$$\theta_5(r, \ell_2) = g \qquad r_s < r < r_0 \tag{25c}$$

$$\theta_5(r, 0) = p \qquad 0 \leqq r < r_0 \tag{25d}$$

$$-\lambda_2\left(\frac{\partial\theta_5}{\partial r}\right)_{r=r_0} = h\{\theta_5(r_0, Z) - a\} \qquad 0 < Z < \ell_2 \tag{25e}$$

$$Q_5 = 2\pi\lambda_2 \int_0^{r_s} r\left(\frac{\partial\theta_5}{\partial Z}\right)_{Z=\ell_2} dr \tag{25f}$$

By the same procedure as before mentioned, the solution of heat flow, Q_5, is to be described by the following summation (refer to Fig. 3*b*)

$$Q_5 = Q_6 - gQ_7 - aQ_8 - pQ_9 \tag{26}$$

where Q_6, Q_7, and Q_9 are the solution of the following three sets of equations

$$\frac{\partial^2\theta_6}{\partial r^2} + \frac{1}{r}\frac{\partial\theta_6}{\partial r} + \frac{\partial^2\theta_6}{\partial Z^2} = 0 \qquad \begin{matrix} 0 < Z < \ell_2 \\ 0 \leqq r < r_0 \end{matrix} \tag{27a}$$

$$\theta_6(r, \ell_2) = 1 \qquad 0 \leqq r < r_s \tag{27b}$$

$$\theta_6(r, \ell_2) = 0 \qquad r_s < r < r_0 \tag{27c}$$

$$\theta_6(r, 0) = 0 \qquad 0 \leqq r < r_0 \tag{27d}$$

$$-\lambda_2\left(\frac{\partial\theta}{\partial r}\right)_{r=r_0} = h\,\theta_6(r_0, Z) \qquad 0 < Z < \ell_2 \tag{27e}$$

$$Q_6 = 2\pi\lambda_2 \int_0^{r_s} r\left(\frac{\partial\theta_6}{\partial Z}\right)_{Z=\ell_2} dr \tag{27f}$$

$$\frac{\partial^2\theta_7}{\partial r^2} + \frac{1}{r}\frac{\partial\theta_7}{\partial r} + \frac{\partial^2\theta_7}{\partial Z^2} = 0 \qquad \begin{matrix} 0 < Z < \ell_2 \\ 0 \leqq r < r_0 \end{matrix} \tag{28a}$$

$$\theta_7(r, \ell_2) = 0 \qquad 0 \leqq r < r_s \tag{28b}$$

$$\theta_7(r, \ell_2) = 1 \qquad r_s < r < r_0 \tag{28c}$$

$$\theta_7(r, 0) = 0 \qquad r \leqq r < r_0 \tag{28d}$$

$$-\lambda_2\left(\frac{\partial\theta_7}{\partial r}\right)_{r=r_0} = h\,\theta_7(r_0, Z) \qquad 0 < Z < \ell_2 \tag{28e}$$

$$-Q_7 = 2\pi\lambda_2 \int_0^{r_s} r\left(\frac{\partial\theta_7}{\partial Z}\right)_{Z=\ell_2} dr \tag{28f}$$

$$\frac{\partial^2\theta_9}{\partial r^2} + \frac{1}{r}\frac{\partial\theta_9}{\partial r} + \frac{\partial^2\theta_9}{\partial Z^2} = 0 \qquad \begin{matrix} 0 < Z < \ell \\ 0 \leqq r < r_0 \end{matrix} \tag{29a}$$

$$\theta_9(r, \ell_2) = 0 \qquad 0 \leqq r < r_0 \tag{29b}$$

$$\theta_9(r, 0) = 1 \qquad 0 \leqq r < r_0 \tag{29c}$$

$$-\lambda_2\left(\frac{\partial\theta_9}{\partial r}\right)_{r=r_0} = h\,\theta_9(r_0, Z) \qquad 0 < Z < \ell_2 \tag{29d}$$

$$-Q_9 = 2\pi\lambda_2 \int_0^{r_s} r\left(\frac{\partial\theta_9}{\partial Z}\right)_{Z=\ell_2} dr \tag{29e}$$

and Q_8 is the solution of the following equation applying the reciprocity theorem (refer to Fig. 3*b*)

$$Q_8 = q_2 = -2\pi\lambda_2 r_0 \int_0^{\ell_2} \left(\frac{\partial\theta_6}{\partial r}\right)_{r=r_0} dZ \tag{30}$$

The solutions of θ_6, Q_6, θ_7, Q_7, θ_9, Q_9, and Q_8 are obtained as follows

$$\theta_6(r, Z) = 2\frac{r_s}{r_0}\sum_{k=1}^{\infty}\frac{n_k}{n_k^2 + C_2^2}\frac{J_1(n_kR_s)}{[J_0(n_k)]^2}\frac{\sinh(n_kZ/r_0)J_0(n_kr/r_0)}{\sinh(n_kL_2)} \tag{31}$$

$$Q_6 = 4\pi\lambda_2 r_s R_s X_2 \tag{32}$$

$$\theta_7(r, Z) = 2\frac{1}{r_0}\sum_{k=1}^{\infty}\frac{n_k}{n_k^2 + C_2^2} \cdot \frac{\{r_0J_1(n_k) - r_sJ_1(n_kR_s)\}}{[J_0(n_k)]^2\sinh(n_kL_2)}\sinh(n_kZ/r_0)J_0(n_kr/r_0) \tag{33}$$

$$Q_7 = 4\pi\lambda_2 r_s(R_sX_2 - Y_2) \tag{34}$$

$$\theta_9(r, \ell - Z) = 2\sum_{k=1}^{\infty}\frac{n_k}{n_k^2 + C_2^2}\frac{J_1(n_k)}{[J_0(n_k)]^2}\frac{\sinh(n_kZ/r_0)J_0(n_kr/r_0)}{\sinh(n_kL_2)} \tag{35}$$

$$Q_9 = 4\pi\lambda_2 r_s W \tag{36}$$

$$Q_8 = 4\pi\lambda_2 r_s Z_2 \tag{37}$$

where $\{n_k\}$ are the roots of

$$n_kJ_1(n_k) - C_2J_0(n_k) = 0 \tag{38}$$

and

$$L_2 = \ell_2/r_o \tag{39}$$

$$\eta = \lambda_2/\lambda_1 \tag{40}$$

$$C_2 = r_0h/\lambda_2 = C_1/\eta \tag{41}$$

$$X_2 = \sum_{k=1}^{\infty}\frac{n_k}{n_k^2 + C_2^2}\left\{\frac{J_1(n_kR_s)}{J_0(n_k)}\right\}^2\frac{1}{\tanh(n_kL_2)} \tag{42}$$

$$Y_2 = \sum_{k=1}^{\infty}\frac{n_k}{n_k^2 + C_2^2}\frac{J_1(n_k)J_1(n_kR_s)}{[J_0(n_k)]^2}\frac{1}{\tanh(n_kL_2)} \tag{43}$$

$$Z_2 = \sum_{k=1}^{\infty}\frac{n_k}{n_k^2 + C_2^2}\frac{J_1(n_k)J_1(n_kR_s)}{[J_0(n_k)]^2}\tanh(n_kL_2/2) \tag{44}$$

$$W = \sum_{k=1}^{\infty} \frac{n_k}{n_k^2 + C_2^2} \frac{J_1(n_k)J_1(n_k R_s)}{[J_0(n_k)]^2} \frac{1}{\sinh(n_k L_2)} \tag{45}$$

Then, substituting Eqs 32, 34, 36, and 37 into Eq 26, the solution of Q_5 is

$$Q_5 = 4\pi\lambda_2 r_s[R_s X_2(1-g) + gY_2 - aZ_2 - pW] \tag{46}$$

According to the assumptions (f) and (g), the direct error heat flows, q_g and q_p, shown in Fig. 2c, may be expressed as follows

$$q_g = \phi_g(1-g) \tag{47}$$

$$q_p = \phi_p(1-p) \tag{48}$$

If $\bar{Q}$ represents the rate of total heat input to HP under the conditions shown in Figs. 2a, b, and c

$$\bar{Q} = Q_1 + Q_5 + q_g + q_p \tag{49}$$

Substituting Eqs 24, 46, 47, and 48 into Eq 49

$$\bar{Q} = 4\pi\lambda_1 r_s\{(R_s X_1 + \eta R_s X_2)(1-g) + (Y_1 + \eta Y_2)g - (Z_1 + \eta Z_2)a - \eta Wp\} + \phi_g(1-g) + \phi_p(1-p) \tag{50}$$

Expression of Errors

If Q_0 represents the rate of one-directional heat flow across the specimen over the HP surface, Q_0 is given by

$$Q_0 = \pi\lambda_1 r_s^2/\ell_1 = \pi\lambda_1 r_s \frac{R_s}{L_1} \tag{51}$$

if λ_s represents the measured thermal conductivity of the specimen, and if λ_s is expressed by

$$\lambda_s = \frac{L_1}{\pi r_s R_s} \bar{Q} \tag{52}$$

a total error $\bar{\epsilon}_T$ is given by

$$\bar{\epsilon}_T = \frac{\lambda_s - \lambda_1}{\lambda_1} = \frac{\bar{Q} - Q_0}{Q_0} \tag{53}$$

Substituting Eqs 50 and 51 into Eq 53

$$\overline{\epsilon}_T = 4\frac{L_1}{R_s}[(R_sX_1 + \eta R_sX_2)(1-g) + (Y_1 + \eta Y_2)g - (Z_1 + \eta Z_2)a - \eta Wp] + \frac{\phi_g(1-g)}{Q_0} + \frac{\phi_p(1-p)}{Q_0} - 1 \quad (54)$$

If the $\overline{\epsilon}_g$, $\overline{\epsilon}_p$, and $\overline{\epsilon}_a$ represent the errors caused by g, p, and a, respectively

$$\overline{\epsilon}_T = \overline{\epsilon}_g + \overline{\epsilon}_p + \overline{\epsilon}_a \quad (55)$$

Substituting $g = p = 1$ into Eq 54 $\overline{\epsilon}_a$ is separated from $\overline{\epsilon}_T$

$$\overline{\epsilon}_a = 4\frac{L_1}{R_s}\{Y_1 - Z_1a + \eta(Y_2 - Z_2a - W)\} - 1 \quad (56)$$

Since $Y_2 - Z_2 - W = 0$ is derived from Eqs 43, 44, and 45, Eq 56 is transformed as follows

$$\overline{\epsilon}_a = \epsilon_{a1} + \epsilon_{a2} = A_1(A_0 - a) + A_2(1-a) \quad (57)$$

where

$$A_0 = \left(Y_1 - \frac{R_s}{4L_1}\right)/Z_1 \quad (58)$$

$$A_1 = 4\frac{L_1}{R_s}Z_1 \quad (59)$$

$$A_2 = 4\frac{L_1}{R_s}Z_2\eta \quad (60)$$

Here, as a part of $\overline{\epsilon}_a$, ϵ_{a1} corresponds to the error caused by the error heat flow in the specimen and ϵ_{a2} corresponds to the error caused by the error heat flow in the insulation between HP and GP (refer to Fig. 4).

Subtracting ϵ_a (Eq 57 through Eq 60) from $\overline{\epsilon}_T$ (Eq 54) and letting $p = 1$, $\overline{\epsilon}_g$ has been obtained as follows

$$\overline{\epsilon}_g = \epsilon_{g1} + \epsilon_{g2} + \epsilon_{g3} = (G_1 + G_2 + G_3)(1-g) \quad (61)$$

$$G_1 = 4\frac{L_1}{R_s}(R_sX_1 - Y_1) \quad (62)$$

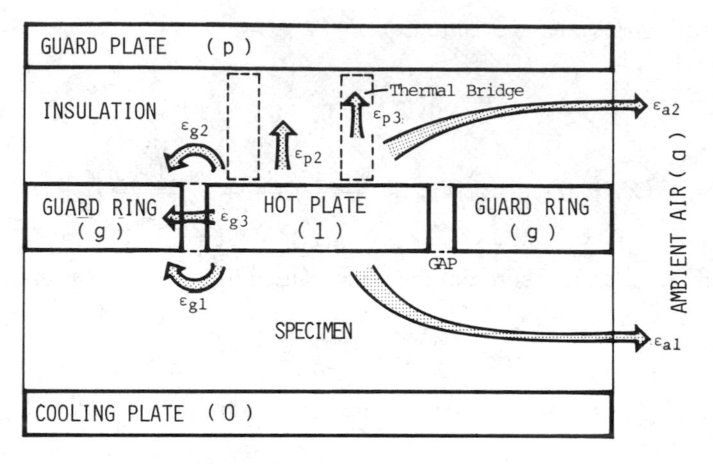

FIG. 4—*Classification of error heat flows.*

$$G_2 = 4\frac{L_1}{R_s}\eta(R_s X_2 - Y_2) \tag{63}$$

$$G_3 = \phi_g \cdot \frac{L_1}{R_s} \cdot \frac{1}{\pi\lambda_1} \cdot \frac{1}{r_s} = \phi_g/Q_0 \tag{64}$$

Here, as a part of $\overline{\epsilon}_g$, ϵ_{g1} corresponds to the error caused by the error heat flow in the specimen, and ϵ_{g2} corresponds to the error caused by the error heat flow in the insulation between HP and GP. ϵ_{g3} is the error caused by the direct error heat flow across the gap (refer to Fig. 4).

In the same way, subtracting ϵ_a from $\overline{\epsilon}_T$ and letting $g = 1$, $\overline{\epsilon}_p$ has been obtained as follows

$$\overline{\epsilon}_p = \epsilon_{p2} + \epsilon_{p3} = (P_2 + P_3)(1 - P) \tag{65}$$

$$P_2 = 4\frac{L_1}{R_s}\eta W \tag{66}$$

$$P_3 = \phi_p \frac{L_1}{R_s} \frac{1}{\pi\lambda_1} \frac{1}{r_s} = \phi_p/Q_0 \tag{67}$$

Here, as a part of $\overline{\epsilon}_p$, ϵ_{p2}, corresponds to the error caused by the error heat flow in the insulation between HP and GP. The error ϵ_{p3} is caused by the direct heat transfer through the thermal bridge between HP and GP (refer to Fig. 4).

From the view-point of the "system," a circular and one side heat flow GHP apparatus is considered to be a system in which the input is temperatures *g*, *p*, and a, and the output is errors $\bar{\epsilon}_g$, $\bar{\epsilon}_p$, and $\bar{\epsilon}_a$. The system parameters that directly connect inputs with outputs are A_0, A_1, A_2, G_1, G_2, G_3, P_2, and P_3. And those system parameters consist of seven nondimensional parameters R_s, L_1, L_2, η, C_1, ϕ_g/Q_0, and ϕ_p/Q_0. Among them, R_s, L_1, and L_2 are decided only by the apparatus dimension, but the others depend on the thermal conductivity and thickness of the specimen and the magnitude of the edge insulation.

In this paper, each of the ϕ_g and ϕ_p is assumed constant for an individual apparatus and considered to be determined from experimental data. For instance, ϕ_g is obtained by subtracting $(\epsilon_{g1} + \epsilon_{g2})$ that is calculated by analytical solution from $\bar{\epsilon}_g$ that is determined by experimental data. The applicability of the model assumed in this paper will be checked by the stability of ϕ_g and ϕ_p derived from the previously mentioned procedures, changing the values of λ_1, ℓ_1, and C_1.

All of the aforementioned formulae are applicable to the circular and two specimen type apparatus, providing the following relationships hold

$$\bar{\epsilon}_g = 2\epsilon_{g1} + \epsilon_{g3} \tag{68}$$

$$\bar{\epsilon}_p = 0 \tag{69}$$

$$\bar{\epsilon}_a = 2\epsilon_{a1} + \epsilon_{a2} \tag{70}$$

Example of Calculation

The following results of G_1, G_2, P_2, and ϵ_2 are calculated using a computer for the BRI apparatus $r_0 = 0.355$, $r_s = 0.15$, $\ell_2 = 0.02$ mm, and $\eta = 1$. The infinite series in the formulae that converges slowly has been calculated up to its 5000 terms.

Calculated G_1 and G_2 shown in Fig. 5 are nearly proportional to the thickness of the specimen and the effects of edge insulation are negligibly small. Calculated G_1 is equal to that of previous report [7] (symbol of $G(L_1, R_s)$ was used in that report). However, in the previous report [7], the term of G_2 was neglected since the heat flow in the insulation between HP and GP was assumed only one-directional heat flow. This mistake has produced an overestimate of ϕ_g in the previous report [7].

The calculated results were $P_2 \simeq \eta(\ell_1/\ell_2)$ and $A_2 = 0$ for $1 < C_1 < 60$. The formula $P_2 = \eta(\ell_1/\ell_2)$ was derived on the assumption of the one-directional heat flow between HP and GP, and this assumption meant $A_2 = 0$ at the same time. Therefore, for the BRI apparatus, the error heat flow which depends on the difference in temperature between HP and GP was able to be estimated easily assuming one-directional heat flow.

The calculated results of ϵ_{a1} are shown in Fig. 6. The figure also indicates that the calculated results of the previous report [7] by alternate long and short

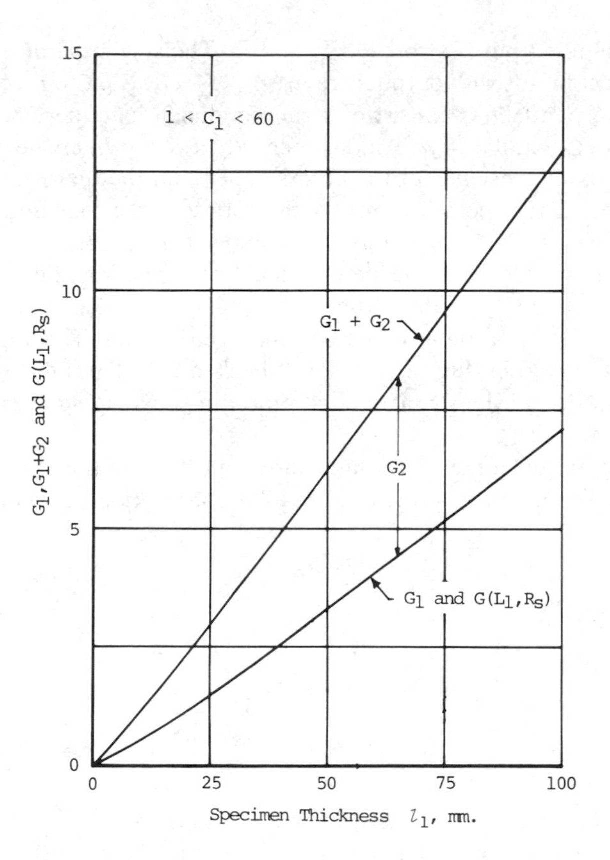

FIG. 5—*Calculated* G_l, $G(L_l, R_s)$ *and* $G_l + G_2$ *versus specimen thickness.* *G_l *may correspond to* $G(L_l, R_s)$ *in the previous model [7].*

dash lines. In the previous report [7], the error caused by edge heat loss was analyzed assuming constant edge temperature, e. The alternate long and short dash lines in Fig. 6 indicate ϵ_{a1} in case of $e = 0$ and $e = 1$, and these correspond to $a = 0$ and $a = 1$, respectively, for $h = \infty$ in this paper. The figure shows that the magnitude of error, caused by edge heat losses through a specimen at a fixed value of a and edge insulation, is small if the thickness of specimen ℓ_1 is less than 50 mm, but it becomes considerably large if ℓ_1 exceeds 50 mm. The smaller the equivalent edge insulation d_1 is, the larger the error is. Figure 6 also shows that it is very difficult to minimize the error due to edge heat losses by the edge insulation only if the thickness of the specimen exceeds 50 mm. Though Eqs 57 through 60 show that ϵ_{a1} is not zero when $p = 1/2$, Fig. 6 indicates that the assumption of $A_0 \simeq 1/2$ is acceptable if the thickness of the specimen is less than 75 mm.

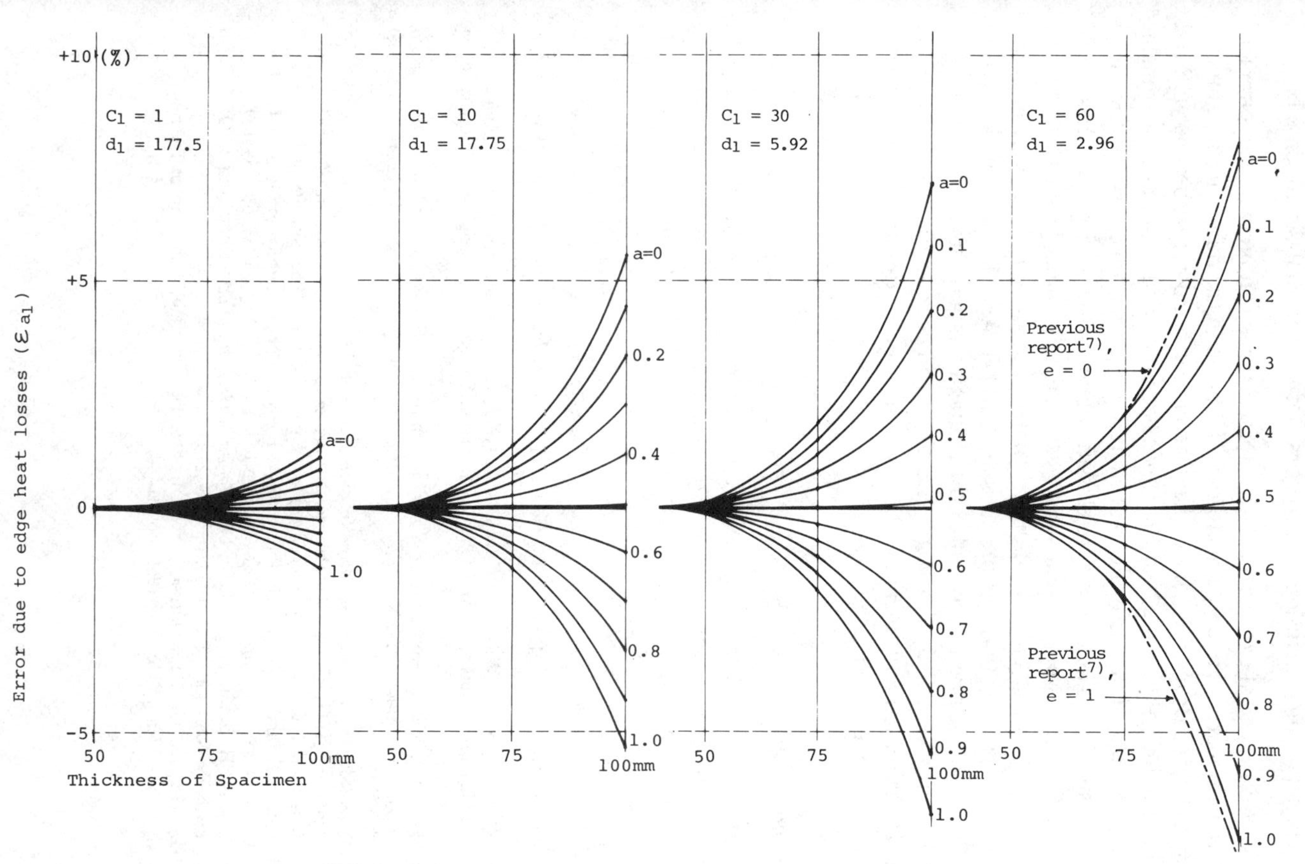

FIG. 6—*Calculated* ϵ_{pl} *versus specimen thickness for different values of* a.

Conclusions

The following conclusions may be drawn from the analysis presented:

1. An analytical expression for the total error in conductivity measurement by the circular and one side heat flow GHP, due to the difference in temperature between HP and GR, due to the difference in temperature between HP and GP, and due to the edge heat losses has been obtained, assuming two-dimensional radial heat flow in cylindrical coordinates on the edge surfaces and in the edge insulation. The errors were classified into the ones caused by these three causes, and they were also classified according to the heat flow paths.
2. Results of calculation indicate that the analysis presented is more rational and realistic than the previous analysis [7]. And the calculated results of ϵ_{a1} also indicate that the control of the ambient temperature is important.
3. Experimental data are still necessary to check the applicability of the model presented and to enable the assessment of maximum error and correction of the measured value.

Acknowledgment

The author wishes to express deepest appreciation to M. Uezono and Y. Miyake for their valuable advices; to Y. Ishiyama and Y. Kawai for their help to translate the paper into English; and to S. Endo for typing the manuscript.

References

[1] Somers, E. V. and Cyphers, J. A., *Review of Scientific Instruments*, Vol. 22, 1951, pp. 583-586.
[2] Woodside, W. and Wilson, A. G. in *Thermal Conductivity Measurements and Applications of Thermal Insulations, ASTM STP 217*, American Society for Testing and Materials, Philadelphia, 1957, pp. 32-48.
[3] Woodside, W. in Symposium on *Thermal Conductivity Measurements and Applications of Thermal Insulation, ASTM STP 217*, American Society for Testing and Materials, Philadelphia, 1957, pp. 49-64.
[4] Woodside, W., *Review of Scientific Instruments*, Vol. 28, 1957, pp. 1033-1037.
[5] Donaldson, I. G., *Quarterly of Applied Mathematics*, Vol. XIX, 1961, pp. 205-219.
[6] Donaldson, I. G., *British Journal of Applied Physics*, Vol. 13, 1962, pp. 598-602.
[7] Eguchi, K. and Miyake, Y. in *Proceedings*, Fourth International Symposium on The Use of Computers for Environmental Engineering Related to Buildings, 1983, pp. 134-139.
[8] Carslaw, H. S. and Jaeger, J. C., *Conduction of Heat in Solids*, second edition, Oxford University Press, Oxford, UK, 1959, pp. 218-219.
[9] Eguchi, K., "On the Reciprocity Theorem of Heat Conduction," Summary of Technical Paper of 37th AIJ Kanto Meeting, 1966, pp. 81-84.

Brian Rennex[1]

Summary of Error Analysis for the National Bureau of Standards 1016-mm Guarded Hot Plate and Considerations Regarding Systematic Error for the Heat Flow Meter Apparatus

REFERENCE: Rennex, B., **"Summary of Error Analysis for the National Bureau of Standards 1016-mm Guarded Hot Plate and Considerations Regarding Systematic Error for the Heat Flow Meter Apparatus,"** *Guarded Hot Plate and Heat Flow Meter Methodology, ASTM STP 879*, C. J. Shirtliffe and R. P. Tye, Eds., American Society for Testing and Materials, Philadelphia, 1985, pp. 69–85.

ABSTRACT: A summary is given of the error analysis of the National Bureau of Standards (NBS) 1016-mm guarded hot plate (GHP) apparatus, with emphasis given to new ideas or techniques. The report discusses the following items: a brief introduction to the principle of measurement, a rationale for the method of summing of uncertainties, a list of individual uncertainties, a description of the innovative plate temperature control method, and a derivation of the error due to systematic uncertainty for the heat flow meter apparatus.

KEY WORDS: apparent thermal conductivity, error analysis, guarded hot plate, heat flow meter, thermal insulation, thermal resistance

This paper was presented at a forum on the state of the art in guarded hot plate (GHP) and heat flow meter (HFM) apparatus. Correspondingly, it will discuss in some detail techniques or ideas that are new or controversial. More complete descriptions of the design and construction of the National Bureau of Standards (NBS) 1016-mm guarded-hot-plate apparatus are given in Ref *1*, and a more detailed version of the error analysis is given in Ref *2*. First, there is a description of the principle of measurement followed by a discussion

[1]Physicist, National Bureau of Standards, Washington, DC 20234.

of how to add the individual uncertainty contributions. Then a table summarizing the individual contributions to the overall uncertainty is presented. Next the methods to estimate the individual uncertainties are described. There is a discussion of the innovative plate temperature control design and capability of the NBS GHP apparatus. Finally, there is a brief discussion of systematic error for HFM apparatus.

Apparatus Description

The GHP measures the heat flow through a sample for a particular boundary condition. Figure 1 shows the basic features of the apparatus in a two-sided configuration. There is a resistance heater in the meter area, A, of the hot plate. The power, Q_m, generated by this heater is metered. In order to ensure the most accurate and repeatable characterization of a sample, it is necessary that this heat flow from the meter area straight across to the cold plates—that is, that its direction be in one dimension. The guard area of the hot plate is maintained at the same temperature as the meter area to prevent lateral heat flow from the meter area. Assuming that the meter power, Q_m, is split between the two sides, the thermal resistance, R, to the heat flow across the meter area is calculated by

$$R = \Delta TA/Q \tag{1}$$

That is, R is defined in terms of the basic quantities Q_m and A, and in terms of the boundary conditions, namely, the hot plate and the two cold plate temperatures. The term ΔT is the average of the temperature difference between the hot plate and the two cold plates. An additional parameter is the speci-

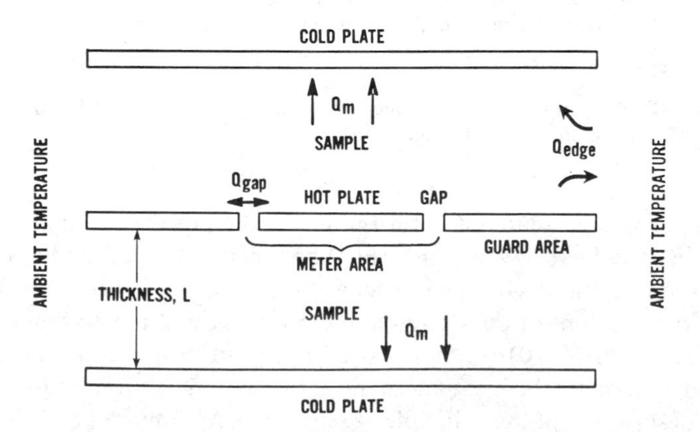

FIG. 1—*Principle of measurement for a guarded hot plate.*

men thickness, L, averaged over both sides. The apparent thermal conductivity, λ, is defined by

$$\lambda = L/R = QL/\Delta TA \tag{2}$$

Note that each of these parameters represents an average over the meter area and over the two sides. Also, the calculated quantities R or λ refer to a steady-state test condition. The test is in steady state at that time at which subsequent values of R (or λ) vary randomly about the mean value. If one of the cold plate temperatures is approximately equal to the hot plate temperature, the apparatus can be run in a one-sided mode.

Propagation of Errors

The apparent thermal conductivity and the thermal resistance of an insulation sample are calculated quantities based on several measured parameters. Each of these measured parameters has an associated uncertainty. In turn, this uncertainty has a random and a systematic (or constant) part. It is possible to estimate the uncertainty of each parameter by an independent test. For example, the apparatus thickness readout can be compared with another thickness gage placed between the plates. It is less straight forward to estimate the overall uncertainty of the calculated quantity, λ. This is because there is usually not sufficient information on the breakdown between the random and systematic parts of the uncertainty for each individual parameter. In principle, it is possible to gather this information, but in practice it would be too time consuming.

A more simple and practical approach was used in this error analysis. The individual parameters, such as the thickness or temperature distribution over the meter area, were measured with an independent detector under test conditions. A comparison of the apparatus readout with these independent measured values made possible the estimate of an upper bound on the total uncertainty for each parameter. Since there is not sufficient information to assure that the measured values are distributed randomly about a "true" mean value, the upper bounds for each individual parameter are simply added to arrive at the overall uncertainty. This is different from an alternative approach to calculate the total uncertainty as the square root of the sum of the squares of the individual uncertainties. This "upper bound" approach results in a somewhat larger estimate of the overall uncertainty (by as much as 30%), but it avoids the need to make an inordinate number of checkup measurements to assure that there are no outlier values. This more conservative approach is thought by the author to be appropriate for a national insulation standards laboratory.

The following philosophy was used with regard to the estimate of upper bounds. Even if an uncertainty might have been expected to be smaller, based on theoretical considerations and manufacturer specifications, the uncertainty value actually used was that of the detector making the independent check. For example, the plate temperature might very well be known within 5 or 10 mK. The uncertainty value actually used, of 22 mK, was associated with the thermopile used to independently check the plate temperature.

It is possible to determine the overall random uncertainty, or repeatability with repeated measurements on the same sample with the same apparatus. The measured short-term reproducibility for the NBS 1016-mm GHP apparatus at a thickness of 100 mm was estimated by the range of values, which was within 0.1%.

Generally speaking, the data on a single test have two parts. The transient part at the beginning of the test shows a monotonically increasing or decreasing curve. When there is no monotonic trend, the steady-state condition has been achieved. There is still a scatter of data points due mostly to the cycling of the bath temperatures. The scatterband is about 1 mK for the hot plate temperature, 6 mK for the cold plate temperatures, and 1 mW for the power. The scatter in the calculated λ-value can be about 0.01% for a two-sided, 1-in. sample, and 0.03% for a two-sided, 6-in. sample. The mean value, of course, is known even better. Clearly, the scatter in the data points, after the steady-state condition is attained, is negligible compared to the estimated systematic errors in the λ-value.

Discussion of Individual Contributions to the Apparatus Uncertainty

Overview

The calculated apparent thermal conductivity, λ, depends on eight measured parameters. In addition to the parameters in Eq 2, there are the gap voltage and the ambient temperature, which involve the heat flow from the meter area—through the gap and to the sample edge. These relate to the assumption that the heat generated in the meter area flows one-dimensionally across the sample. The salient features of the methods to estimate the uncertainty will be discussed here (a detailed discussion is given in Ref *2*.)

Before looking at the individual uncertainties, please look at the overall picture as portrayed in Table 1. This summary of uncertainties is for the case of compressible insulation samples at a mean temperature difference of 27 K. Note that a thickness of 25 mm, the temperature and thickness uncertainties are largest. At 150 mm, the temperature is the only large uncertainty. At 300 mm, the edge uncertainty is dominant. The gap-voltage uncertainty was kept small, even at 300 mm, by using an 18-stage gap thermopile, low-thermal wiring and a highly accurate voltmeter. The overall uncertainty value is ~ 1/3% up to a 150 mm thickness. At 300 mm it should be possible to decrease the uncertainty with further edge studies.

TABLE 1—*Percentage estimate of uncertainties in the measured apparent thermal conductivity for the NBS guarded hot plate.*

Quantitative Value[a]	Thickness: 25 mm (1 in.)	75 mm (3 in.)	150 mm (6 in.)	300 mm (12 in.)
	Percent Uncertainty			
Area (12 μm or 0.5 mil in radius)	0.01	0.01	0.01	0.01
Thickness (25 μm or 1.0 mil)	0.1	0.03	0.02	0.01
Meter power	0.04	0.04	0.04	0.04
Meter resistive device (0.4 mW)	0.00	0.01	0.02	0.04
Gap heat flow (0.3 mW, or 0.5 μV in gap voltage)	0.00	0.01	0.02	0.03
Edge heat flow	0.00	0.00	0.00	0.50
Hot, cold-plate temperature difference (44 mK)	0.16	0.16	0.16	0.16
Total	0.31	0.26	0.27	0.79

[a]These values are for compressible, low-density (9.6 kg/m^3), glass-fiber insulation measured in the two-sided mode with a plate temperature difference of 28 K. Uncertainty values of less than 0.01% are reported as zero.

Meter Area

The meter area used corresponds to the radius at the geometric center of the gap, which is the square root average of the squares of the inner and outer radii of the gap. The uncertainty in this meter area is not more than 0.01%.

Thickness

The parameter L refers to the average specimen thickness over the meter area. For low-density or compressible specimens this is best estimated by the value of the average plate spacing over the meter area. Precaution should be taken to compress such a sample slightly to prevent voids between the specimen and the plates. For rigid specimens, there are two possible ways to estimate the test thickness. First, if the specimen surfaces and the plates are sufficiently flat and parallel, then the plate spacing is a good estimate. Second, if any of these surfaces are irregular, then the specimen thickness itself should be measured—with a caliper, for example. An effort should be made to estimate of the thermal resistance of the small air gaps at the irregular surfaces.

The measurement of the plate spacing is described next. This shows the calibration method to measure the positions of the four outside "corners" of the cold plates relative to a known meter area plate separation. These outside cold-plate positions are measured using four thickness transducers on each plate. These thickness transducers then measure any change in position relative to the initial calibration point. This calibration must be done for each plate orientation and for compressible and rigid samples.

The following is a more detailed discussion of the thickness calibration—referred to as the L-map. Please refer to Fig. 2. Here is shown the basic plate support and measurement system. The hot plate is rigidly and permanently mounted on the four support rods. The cold plates are supported in the center. This point of support has a load cell to measure the force that the sample exerts on the plate. It also has a ball joint so that the plate can tilt to conform to a nonparallel rigid sample. The cold plates are constrained in the radial direction by steel cables attached to four spring loaded bearings which are mounted on stainless-steel rods with a diameter of 51 mm (2 in.). At four points (at 90° intervals) at the edge of each cold plate the positions (perpendicular to the plates) are measured with thickness transducers (referred to as FTT). The manufacturers stated overall accuracy over a displacement of 150 mm (6 in.) is within 10 μm (0.0004 in.) for a single FTT. The mean displacement of the four FTTs on a single cold plate would be then known within a factor of $1/\sqrt{4}$ or 5 μm (0.0002 in.). The FTTs are mounted on Invar bars to minimize error due to ambient temperature variation. A coefficient of thermal expansion of 10^{-6} m/m K, a length of 0.25 m (10 in.), and a temperature range of 50 K correspond to a length change (or error) of 13 μm (0.5 mil). This contribution to error is avoided by the calibration procedure to be described later, since it is performed at the ambient temperature appropriate to the test.

The calibration procedure is to correlate an absolute measurement of the average meter area spacings with average displacements of the FTTs. The FTTs will then provide any displacements from this original calibration

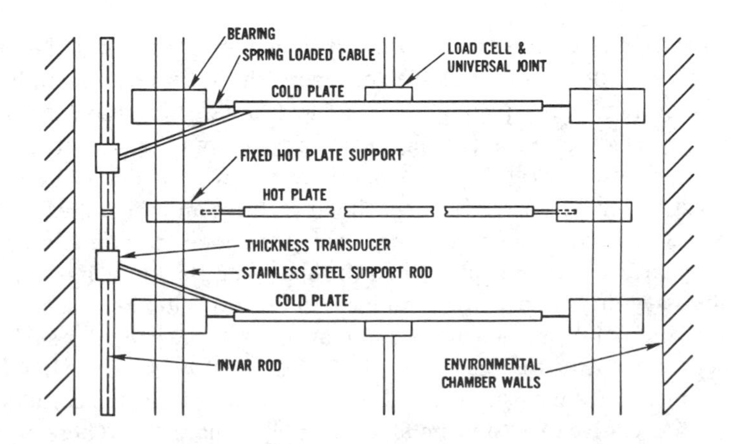

FIG. 2—*Plate support and thickness calibration.*

point, provided the plates remain in the same orientation—that is, that they are not rotated. The absolute measurement of the average meter area spacing is made with portable thickness transducers placed directly over the meter area. This measurement is made with the plates horizontal and vertical, since the plate sag, or deformation, is different in the two cases. These thickness transducers were calibrated with an uncertainty of 5 μm, (0.2 mil), using NBS gage blocks accurate within 0.3 μm (0.01 mil) at 293.15 K (20°C). The uncertainty in the gage block length, due to temperature difference from that of calibration when the gage block was calibrated, is estimated to be 3 μm (0.1 mil). The measured flatness within the meter area is +25 μm (1 mil). In the L-map, the plate spacing is measured at 25 points in the meter area. The repeatability of the average of these 25 points was estimated by the range of values, which was 5 μm (0.2 mil). Thus, the uncertainty is about $1/5 = 1/\sqrt{25}$ times the flatness.

The next step is to measure the difference between the spacing at the center of the meter area and the average spacing over the meter area (the average of the 25 points). The reason is that the calibration procedure can then rely on the measurement at one point at the center, rather than require 25 points. Another 2.5 μm (0.1 mil) value is added due to the transducer uncertainty for the transducer center point measurement.

To understand the final contribution to thickness uncertainty we must look at the problem of plate bowing deformation. The plates are 1 m (40 in.) in diameter. The cold plates are 19 mm (0.75 in.) thick and weigh about 50 kg (110 lb). The hot plate is 16.5 mm (0.65 in.) thick and weighs about 32 kg (70 lb). Referring to Fig. 2, the various forces that act to deform the plate are due to the load cell at the center, the weight of the plates, and the spring-loaded cables at the four corners. As the cold plates open and close, they tilt, which causes a change in deformation. The problem with the bowing deformation is that the FTT readings correspond to four thickness points at the plate edges. The average of these is not necessarily the same as the value at the plate center, when there is a deformation. Repeatability studies compared the FTT thickness values with those of a transducer at the plate center, after the cold plates have been opened and closed back to the original thickness. These gave a repeatability estimated by a range of values of 5 μm (0.2 mil). Note that is is important that the stainless-steel, plate-support rods be straight and parallel to have the best repeatability.

To review, the calibration procedure uses thickness transducers which are calibrated with NBS gage blocks and placed in the center of the meter area, between the hot and cold plate, at the test thickness and temperature. The transducer reading then is used to set the 8 FTTs (4 on each side). The plates are then opened, the samples are put in, and the plates are closed to the test condition. Below is a summary of the uncertainties, Δ_i, that contribute to the thickness uncertainty for compressible low-density specimens.

Source	Δ_i	
	μm	mil
Auxiliary thickness transducer	5	0.2
Thermal expansion of gage block	2.5	0.1
Δ between average and center meter-area spacing	5	0.2
Δ of center position	2.5	0.1
FTT (thickness transducer)	5	0.2
Repeatability after opening and closing cold plates	5	0.2
Total Δ_i	25	1.0

Heat Flow

There are a number of heat flows, designated by Q, involved in the guarded hot plate apparatus (refer to Fig. 1). There is the heat generated by the meter heater, Q_m, and the heat produced by any resistive sensors in the meter area, Q_r. There is the heat flow between the meter area and the guard area, Q_{gap}. There is some net heat flow between the meter area and the ambient through the specimen, called the edge heat flow, Q_{edge}. Q_{edge} depends on the guard and meter widths, on the thickness, on the ambient and plate temperatures, and on the sample λ-value. The goal of a proper measurement procedure is to minimize any systematic error that might result from these heat flows.

The measurement of the meter heater power and the power due to the resistive sensors in the meter area is straightforward, and their combined uncertainty amounts to less than 0.08% (see Table 1).

The measurement of the heat flow across the gap is more difficult, and it involves several precautions. To have the best confidence that there is a zero net heat flow across the gap, it is recommended that a large number, N, of stages be used in the differential thermopile that measures the average temperature difference across the gap. This number, N, is limited by the space limitation for the wires in the gap. Also, small wires of diameter ~0.3 mm (10 mil) should be used. These should not go directly across the gap, but rather should go sideways around the gap. That is, the length between two adjacent thermocouple junctions is about 8 cm. This is done since the thermal resistance of the wire to heat flow across the gap is inversely proportional to the wire length. With these precautions, each thermocouple adds only about 1/1000 of the heat already flowing across the gap through the air. Clearly, it is better to have a larger number of thermocouple stages, N, since the sensitivity of the thermopile increases proportional to N.

Another reason to have a large N is that the temperature difference across the gap is not uniform as a function of position on the gap perimeter. That is, the heat flow will be inwards in some positions, and outwards in other positions. This was, in fact, measured on the NBS GHP apparatus. If a small number of thermocouple stages were used and if they happened to be located where the heat was only flowing inwards, the thermopile voltage would not be indicative of the net gap heat flow. That is, there could be a systematic error. The use of a larger number N minimizes the chance of such an error.

The following is a detailed account of the sources of error in the measurement of the gap-thermopile voltage. Figure 3 shows a graphical representation of the gap measurement situation. As long as it is a null measurement with the meter, gap, and guard temperatures at the same value, there is no readout error in the thermopile signal up to Point A.

At Point A the 0.25 mm (10 mil) Chromel wire is soldered to the 0.25 mm (10 mil) copper wire, which is very pure and, hence, should have a very small thermal signal ($\lesssim 0.1$ μV) coming out through the guard. The order of magnitude of the spurious thermal signal due to a temperature difference between the last Chromel-Constantan junction and the Chromel-copper junction can be estimated as follows. The two Chromel-copper junctions are located in the same location on the guard side of the gap, so an overestimate of the temperature difference between their locations would be 1 m K. The chromium-copper sensitivity is ~20 μV/K. Thus, the spurious signal would be about 0.02 μV, and this is negligible.

The next error source is the thermals generated in the pure copper wire between Points A and B in Fig. 3. If the wire is handled with care, and low-thermal solder is used, these thermals should be less than 0.1 μV, and this is referred to as ΔV_{AB}. A value of 0.2 μV for ΔV_{AB} was used. (If the copper wires are connected at Point A, the actual thermal signal can be measured within the accuracy of the readout device.)

The next error results from any thermals in the reversing switch B or be-

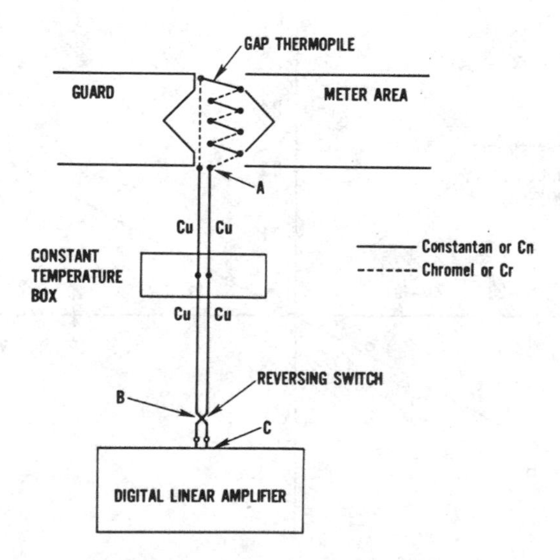

FIG. 3—*Voltage circuit and wire configuration for the gap thermopile.*

tween B and C. The latter error can be eliminated by reversing the signal at B. Any residual thermal signal in the reversing switch can be measured by shorting a copper wire across the switch. This quantity, ΔV_{switch}, is estimated at 0.2 μV. Finally, there is the readout error of the digital linear amplifier (DLA), ΔV_{DLA}. This is estimated to be about 0.1 μV. The sum of ΔV_{DLA}, ΔV_{switch}, and ΔV_{AB} is 0.5 μV, and this is referred to as ΔV_g. Remember that the other contributions to error were estimated to be negligible.

The resulting heat flow uncertainty, ΔQ_g is calculated by $\Delta Q_g = S_g \, \Delta V_g$, where S_g is the gap sensitivity. The term S_g is equal to the slope of the curve of the change in heat flow, Q_g, as the gap thermopile voltage is unbalanced by an amount equal to V_g (Fig. 4), $S_g = Q_g/V_g$.

The quantity S_g has a value of 0.57 mW/μV, so ΔQ_g is estimated to be 0.3 mW. The percent uncertainty due to the gap is calculated as the ratio, $\Delta Q_g/Q_m$. In the two-sided configuration, $Q_m \sim 1$ W at 300 mm (12 in.) for low-density insulation samples. The corresponding percentage uncertainty value is ~0.03%.

Q_{edge}

Looking at the heat flow from the hot-plate meter area, the difference between the ideal case with one-dimensional heat flow and the real case is due to the edge heat flow, Q_{edge}. Referring to Fig. 1, the ambient temperature, T_{amb}, is different from that of either the hot or cold plate. Usually, T_{amb} is con-

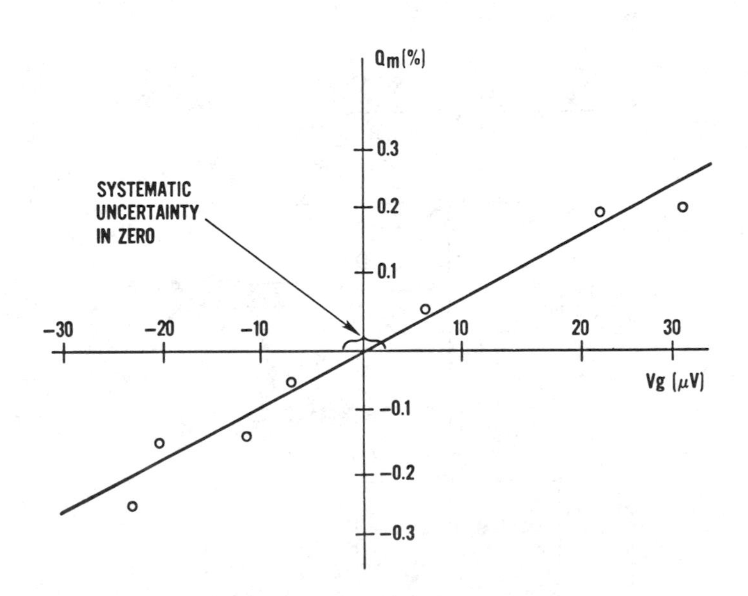

FIG. 4—*Hot-plate, meter-area power,* Q_m, *versus gap voltage.*

trolled equal to the average of the hot and cold plate temperatures, T_{mean}. At first glance, one would expect the net Q_{edge} to be zero when T_{amb} is equal to T_{mean}. This would be true if heat flow were measured midway between the hot and cold plates. In fact, the heat flow is monitored at the surface of the hot plate. Some of the heat from the hot surface is going to the ambient, so, the measured heat flow is greater than what would be the case for one-dimensional heat flow. Therefore, the ambient must be equal to a temperature higher than T_{mean}, for Q_{edge} to be zero.

This leads to the essential difficulty in estimating the edge problems, which is the issue of determining the value of T_{amb} at which Q_{edge} is zero. Figure 5 shows the plot of a theoretically calculated plot [*2*] of the percentage change in λ as the ambient temperature varies. Since λ is proportional to Q, the edge effect can be characterized by λ as well as Q. A value of zero for the ordinate corresponds to the case where λ_{edge} (or Q_{edge}) equals zero. Note that this occurs when T_{amb} is greater than T_{mean}. It is possible to calculate the value of T_{amb} at which Q_{edge} is zero, and one such calculation is given in Ref *2*. The next step is to consider how to check these calculations with experiment.

It is a simple matter to measure the slope of the curve in Fig. 5. It is more difficult to determine the zero crossing of the ordinate—at which (1) Q_{edge} equals zero or, equivalently, at which (2) the Q monitored at the hot plate is equal to what it would be if the heat flow were one-dimensional or, equivalently, at which (3) one is measuring the "true" apparent thermal conductiv-

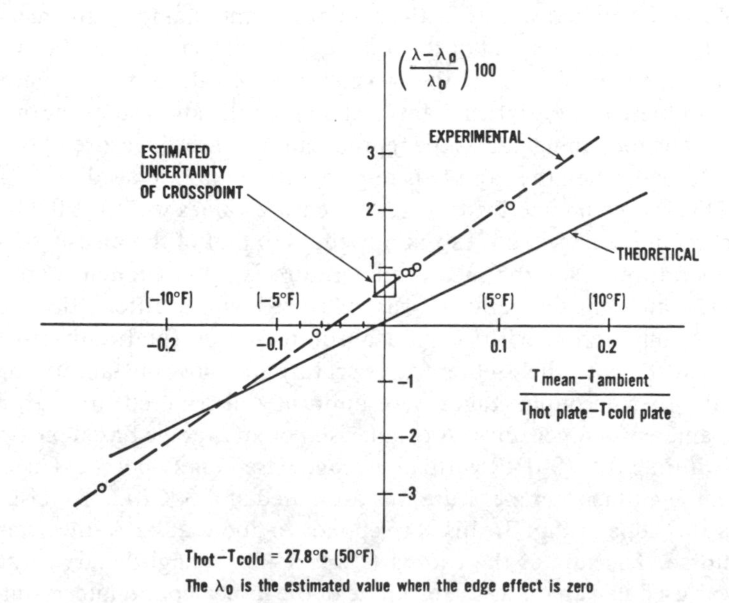

FIG. 5—*The change in apparent thermal conductivity, λ, as a function of ambient temperature.*

ity of the material, λ_0. The only way to determine this zero intercept is to put in a sample with a previously known λ-value. Two 150 mm (6-in.) specimens were measured on the NBS GHP. The edge effect is expected to be negligibly small at this 150 mm (6-in.) thickness, based on theory and experimental measurements of the slope of the curve in Fig. 5 (the value of this slope was less than 0.06 %/K). Therefore, the λ-value of each 150-mm (6-in.) sample is known within the experimental uncertainty at this thickness. Incidentally, these tests were in the one-sided mode, and there was no edge insulation.

The next step in the experiment was to stack the two 150 mm (6-in.) specimens to make a 300 mm (12-in.) sample. The mean temperature was 23.9 K (75°F), and the plate temperature difference ($T_{hot} - T_{cold}$) was 27.8 K (50°F). At this thickness, for the NBS GHP, the edge effect was expected to be significant (~0.4% change in λ-value for a 1 K change). A zero value of the abscissa in Fig. 5 corresponds to the ambient temperature being equal to the mean temperature. A theoretical calculation [2] indicated that the corresponding value of the ordinate would be ~0.07%. Preliminary experimental values were ±0.6 about the expected value. More careful experiments are being carried out to check the theory.

The experimental curve in Fig. 5 can be used to estimate the value of the abscissa (which is proportional to $[T_{mean} - T_{amb}]$) at which the measured λ is equal to λ_0. In this case, the ambient temperature would have to be 1.4 K (2.5°F) more than the mean temperature to measure the true λ-value ($\lambda - \lambda_0 = 0$). The uncertainty, as far as the edge effect is concerned, with which λ_0 is known is estimated as follows. The uncertainty in the measured value of λ_0 is estimated simply as the sum of the uncertainty of the measured value of the abscissa and that of the ordinate. The uncertainty of the ordinate is based on the uncertainties of the two measured λ-values at 150 mm (6-in.). The part of these uncertainties that is systematic will cancel since the ordinate is a ratio. The remaining part is the reproducibility of the measured λ-value at 150 mm (6-in.), when the sample is removed from and replaced in the apparatus. This was estimated by a range of measure values within ±0.2%. The uncertainty in the abscissa is essentially equal to that of the measured ambient temperature, since the mean temperature is known much more accurately. The ambient temperature was measured with a differential thermocouple in conjunction with a good absolute sensor (platinum-resistance thermometer). Since the ambient temperature was not constant around the plates, the thermocouple stages were uniformly distributed to measure the average ambient temperature. A comparison of an average based on 4 points agreed within 0.3 K (0.6°F) with an average based on 24 points. The uncertainty in the ambient temperature was estimated at 0.3 K (0.6°F). Using the value of the slope in Fig. 3, this corresponds to about a 0.2% uncertainty in the ordinate. The sum of the two parts was 0.4%. A slightly larger value of 0.5% was used in Table 1 as the estimate of the upper-bound uncertainty of a measured λ-value at 300 mm (12-in.), due to edge effects.

Temperature Measurement

Overview

In order to calculate the apparent thermal conductivity, λ, via Eq 2, one must measure the plate temperatures. Platinum resistance thermometers (PRTs) and thermocouples are used in this measurement. The PRT is used to determine the absolute temperature at a particular plate location, and differential thermocouples are used to measure the relative temperatures between the PRT and other locations. The discussion of the uncertainties of the average, meter-area, plate-temperature values follows the discussion of the uncertainties of the PRT and thermocouples values.

PRT Circuit Rationale

Platinum resistance thermometers are used to measure the absolute temperature in the meter area of the hot and cold plates. The circuit in Fig. 6 is used to measure the PRT resistance values. The value of the PRT resistance, R_x, is $R_x = (V_x/V_{std})R_{std}$. Note that the digital voltage meter (DVM) voltage uncertainty due to the zero uncertainty is effectively eliminated by taking the ratio. Also, the DVM uncertainty due to linearity is effectively eliminated by matching the values of R_x and R_{std}. Thus, the readout percentage uncertainty of R_x, ΔR_x, is approximately equal to that of R_{std}, which is $\pm 0.001\%$. Since a change of 1 Ω corresponds to a change of about 2 K for these "100 Ω" PRTs, the value of 0.001% corresponds to 1 mΩ or 2 mK.

The point of view discussed in the beginning of this paper was used in the

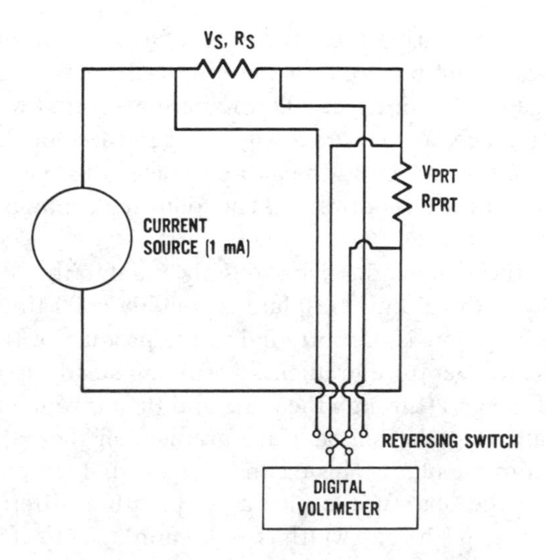

FIG. 6—*Circuit for the platinum-resistance thermometer (PRT) resistance measurement.*

temperature uncertainty estimate—namely, that the actual, *in situ* temperature distribution should be measured with an independent detector and that the uncertainty of this independent detector be the lower bound of the plate temperature uncertainty. Stated in other words, even if theory might provide a rationale that the uncertainty could be much smaller, one should claim only what can be demonstrated with an independent measurement.

A 16-stage thermopile was used to accomplish this independent measurement of the temperatures on the plate surfaces. It was laid flat on the plate surfaces, and it gave results repeatable to within 15 mK. Many measurements on both the hot and cold plates indicated that the temperature at the PRT location in the plates was representative of the average temperature within this value of 15 mK [2].

Summary of Temperature Uncertainties

The sum of the various contributions, ΔT, to the uncertainty in the measurement of the average T over the meter area is as follows.

ΔT due to PRT resistance		2 mK
ΔT due to bath calibration		5 mK
ΔT due to temperature distribution		
Not constant over meter area	(hot plate)	15 mK
	(cold plate)	15 mK
Total	(hot plate)	22 mK
Total	(cold plate)	22 mK

Temperature Jump at the Gap

In the course of measuring the temperature distribution in the hot plate, the following interesting phenomena were noticed. Differences of temperature across the gap of the order of 200 mK were measured when the net gap temperature difference, as measured with the gap thermopile, was less than 2 mK. In some locations around the gap perimeter the heat was flowing inwards and in other locations outwards. The following explanation is given for this observation.

The gap has a thermal conductivity roughly equal to that of air. The thermal conductivity of the aluminum plate is roughly 6000 times greater than that of air. The heat flow is proportional to the product of the thermal conductivity and the temperature gradient. Assuming steady state, then the radial heat flow at the gap (through the plate and then through the gap) is constant. This means that the temperature gradient in the gap is 6000 times greater than that in the plate. This means that a radial temperature gradient of 0.2 mK/cm in the plate would give a temperature difference of roughly 100 mK across the gap (the gap width is ~0.9 mm). In effect, the gap acts as an amplifier of the temperature gradient.

It would be impossible to make a realistic calculation of the angular distribution of the temperature difference across the gap, because of the aforementioned amplification of temperature difference across the gap and because the plate gradients that are being amplified are so small. Fortunately, it is possible with the thermopile to ascertain empirically that the net gap heat flow is zero.

Plate Emittance

The heat flow, and, hence, the thermal resistance, through a sample depends on the emittance of the apparatus plates. In the case of an insulation sample material which has a significant fraction of heat transfer via radiation, the apparatus plate emittances must be taken into account when two different apparatuses are compared. For a 25 mm thick, 10 kg/m^3 density mineral fiber insulation sample, a change in plate emittance value between 0.8 to 0.9 results in a difference in thermal resistance value of 1.2% [*2*]. Note, this effect rapidly becomes negligible for samples of greater thickness or higher density. Thus, if two apparatuses were identical except for this difference in plate emittances and if they measured an identical low-density sample, their results for thermal resistance would differ by 1.2%. For this reason, it is important that the actual hemispherical emittance of the plates be measured. The recommended ASTM standard value is 0.9 at room temperature ($\sim$300 K).

Plate Temperature Control

The control of the plate temperatures was accomplished with feedback circuits. Thermistors were used in these circuits due to their large change in resistance with temperature. These were located in the hot and cold plates. Pairs of thermistors were used to control the temperature across the gap and the outer guard. For the NBS GHP apparatus, it is possible to control the hot plate temperature with a scatter of 1 mK, the cold plate temperatures within 6 mK, and the average temperature difference across the gap within 1 mK. A microprocessor was used in conjunction with the feedback circuit to automatically bring the plate temperatures to a desired value. The use of the previously mentioned thermistor control circuits made it possible to achieve steady-state test conditions within 3 h, even for low-density samples 150 mm thick. This constituted a considerable reduction in test times (by a factor of 5) from what was possible using a thermopile control circuit.

The steady-state values of apparent thermal conductivity are achieved in the constant, hot plate temperature mode, rather than a constant meter-area-heater power mode. This is also a factor in the reduction of test times. Note that the control range of the hot plate temperature and the power is very small. Moreover, the uncertainties in the mean values of these parameters are so small as to be negligible. Thus, the accuracy of the final steady-state value of the apparent thermal conductivity is not compromised by using the constant-temperature mode.

Heat Flow Meter Uncertainty

The HFM apparatus relies on a measurement with a sample of known apparent thermal conductivity. This is then used to obtain a calibration constant for the apparatus, which is, in turn, used to obtain apparent thermal conductivity (λ) values for unknown samples. In this sense it is a relative, rather than an absolute, device. The following derivation shows that it is possible to have an error in the measured λ-values of unknown samples when there is a systematic error in the instrumentation of the HFM apparatus. This is worth noting because it is commonly thought that such errors cancel out since it is a relative device.

For simplicity, let us say that the λ-value is proportional to a single measured parameter, V.

$$\lambda = CV \tag{3}$$

C is the calibration constant. Suppose a calibration specimen of λ-value, λ_C, is placed in the HFM apparatus, and the value V_C' is measured. Then the calibration constant is calculated

$$C' = \lambda_C / V_C' \tag{4}$$

Now, let us postulate that there is a systematic error, s, in the HFM parameter readout, so V_C' is actually equal to $V_C + s$, and C should actually be equal to λ_C / V_C.

Now, the question is, "Will this error cancel out in the calculation of an unknown λ-value?" The answer is yes only when the measurement of the unknown sample results in the same measured value of V_C'. In general, this will not be the case. Let V_u' be the measured value when the unknown sample is measured in the HFM. The assumption is that the same systematic error pertains, so $V_u' = V_u + s$, where V_u is the "true" voltage.

To calculate the percentage error in the calculated λ-value, note that the true value is $\lambda_u = CV_u$. The value actually calculated is $\lambda_u' = C'V_u'$. Some algebra shows that

$$\frac{\lambda_u' - \lambda_u}{\lambda_u} = \frac{s(V_c - V_u)}{(V_c + s)V_u} = \frac{s}{(V_c + s)} \frac{\Delta V}{V_u} \tag{5}$$

Here $V_c - V_u = \Delta V$, the difference in range of the measured parameter, V. For example, if V were the heat flow transducer voltage in the HFM apparatus, two samples of different thermal resistance would result in a different range of values for V.

Looking at the last equation, we can think of the percent error in λ-value as being equal to the product of the percent systematic error, $s/(V_c + s)$, times the percent range change, $\Delta V/V_u$. This derivation shows quantitatively that a particular calibration of a HFM apparatus has a wider range of applicability, ΔV_g, when the systematic errors, s, are smaller. The most notable systematic errors for the HFM apparatus occur in the measurement of thickness and plate temperature and in heat loss or gain to the ambient at the sample edge.

References

[1] Hahn, M. H., Robinson, H. E., and Flynn, D. R. in *Heat Transmissions in Thermal Insulations, ASTM STP 544*, American Society of Testing and Materials, Philadelphia, 1974, pp. 167-192.

[2] B. Rennex, *Journal of Thermal Insulation*. Vol. 7, July 1983, pp. 18-51.

Einar Brendeng[1]

Evaluation of Thermal Conductance of Liquefied Natural Gas Tank Insulation

REFERENCE: Brendeng, E., **"Evaluation of Thermal Conductance of Liquefied Natural Gas Tank Insulation,"** *Guarded Hot Plate and Heat Flow Meter Methodology, ASTM STP 879*, C. J. Shirtliffe, and R. P. Tye, Eds., American Society for Testing and Materials, Philadelphia, 1985, pp. 86–97.

ABSTRACT: The activities in testing of liquefied natural gas (LNG) insulation of The Division of Refrigeration Engineering at the University in Trondheim were initiated by the need for a satisfactory insulation system for the spherical tank LNG carriers of Moss Rosenberg Verft. A large scale guarded hot plate apparatus has been built, with test area 2 by 3 m^2. The temperature on the cold plate may be 0 to −162°C, and on the warm plate 30 to —30°C. Measurements may be made in horizontal convection-free mode, and in vertical convection mode.

A load-bearing evacuated insulation may be a possible solution in the efforts to reduce the boil-off of the tank content. A heat flow meter apparatus for testing evacuated insulations under load has been built, with diameter of the specimen 400 mm, and thickness up to 100 mm.

KEY WORDS: guarded hot plate apparatus, heat flow meter apparatus, liquefied natural gas insulation, evacuated insulation

For some years there has been a considerable interest in the building of liquefied petroleum gas (LPG) and liquefied natural gas (LNG) carriers in Norway. The insulation of tanks for propane is comparatively simple, while the insulation of tanks for ethylene and methane is a demanding task, due to the lower temperatures involved. In order to reduce the boil-off of the tank content, the thickness of the insulation may be increased. Evacuated insulation, however, is a possible solution. The insulation costs are high, and a careful evaluation of the insulating system must be made before the erection. In order to carry out the necessary testing, both a large-scale guarded hot plate

[1]Professor, Division of Refrigeration Engineering, The Norwegian Institute of Technology, Trondheim, Norway.

and an apparatus based on the heat flow meter (HFM) principle have been developed especially for this purpose.

Guarded Hot Plate Apparatus—Design Principle

In designing insulation systems for LNG-temperatures, due respect should be paid to the effects of the different thermal contractions of the tank and the insulation. If the insulation is not strong enough to withstand the stresses that arise in a monolithic construction, some type of expansion system must be used. Some of these systems have been the cause of convection currents within the insulation, resulting in severe increases in the heat flow. Any test equipment must permit accurate measurements of the influence of convection, the thermal stresses should be equivalent to the stresses in the actual installation, and the size must be such that serious scale effects do not occur.

Figure 1 shows the design principle for the large-scale guarded hot plate apparatus. The test cavity is surrounded by a perimeter insulation, impermeable to air currents. If convection occurs in the test cavity, the isotherms are distorted, and the heat flow in the convection-free perimeter is also affected. Thus, the influence of convection is not limited only to the test cavity, and the total convection influence is defined as the sum of the convection influence in the test cavity and in the perimeter insulation.

Calculation of the Apparent Thermal Conductivity

The evaluation of the heat leakage through the perimeter insulation is made with the apparatus in horizontal convection-free position, with the test

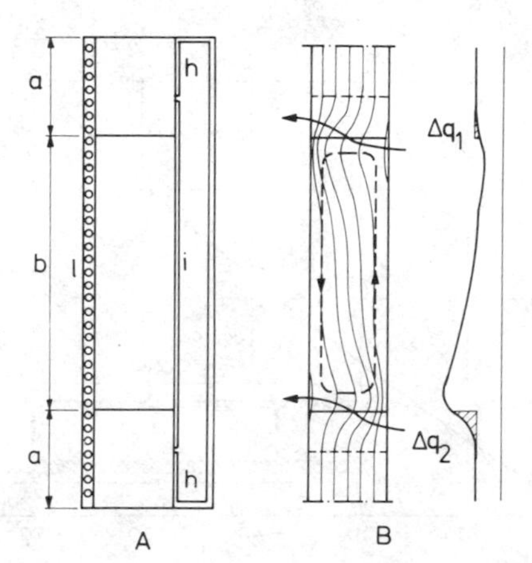

FIG. 1—(*A*) *Principle for large scale guarded hot plate apparatus.* (a) *convection-free perimeter insulation,* (b) *test cavity,* (i) *metering plate,* (h) *guard ring, and* (l) *cold plate.* (*B*) *Convection current, isotherms and heat flow distribution.* ($\Delta q_2 - \Delta q_l$) *is increase in heat flow through perimeter insulation due to distortion of isotherms, caused by convection currents.*

cavity packed with high-density mineral wool. The thermal conductivity of this material was measured in a conventional guarded hot plate apparatus. These measurements are carried out at the same temperature difference and mean temperatures for the perimeter insulation as in the actual tests. The total heat flow Q_{tc} is measured, and the heat leakage Q_p through the perimeter insulation if found by subtraction of the heat flow Q_{ic} through the test cavity

$$Q_{ic} = (\lambda_{ic}/d) \cdot F \cdot \Delta t_{ic}$$

$$Q_p = Q_{tc} - Q_{ic}$$

where

λ_{ic} = thermal conductivity of calibration insulation, W/(m · K),
F = area of test cavity, m^2,
d = thickness of test cavity, m,
Δt_{ic} = temperature difference across test cavity, K.

Figure 2 shows the thermal conductance of the perimeter insulation of extruded polystyrene as a function of the mean temperature.

The apparent thermal conductivity of the insulation to be tested is found by measurements of the total heat flow Q_t and subtraction of the calculated heat leakage Q_p through the perimeter insulation

$$\lambda_i = (Q_t - Q_p) \cdot d/(F \cdot \Delta t_i)$$

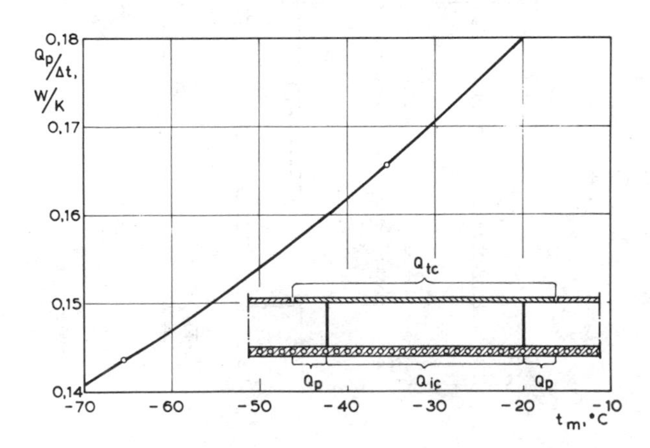

FIG. 2—*Thermal conductance* Qp/Δt *of perimeter insulation, dependent on mean temperature* t_m.

where

λ_i = apparent thermal conductivity for test insulation, W/(m K), and
Δt_i = temperature difference across test insulation, K.

The influence of convection is found by comparison of the apparent thermal conductivity measured in horizontal and vertical positions.

Design Details

The general arrangement of the apparatus is shown in Fig. 3. The apparatus consists of three main parts, the cold plate assembly, the perimeter insulation with the test panel, and the warm plate assembly. The insulation panel under test is installed in the perimeter insulation and kept between the warm and cold plate assemblies by means of hydraulic jacks. Outmost care must be shown in the installation to avoid false air currents.

Figure 4 shows the design of the different parts of the apparatus.

The cold plate is cooled by means of propane circulated through tubes in the plate by a hermetic pump. Figure 5 shows the arrangement of the tubes.

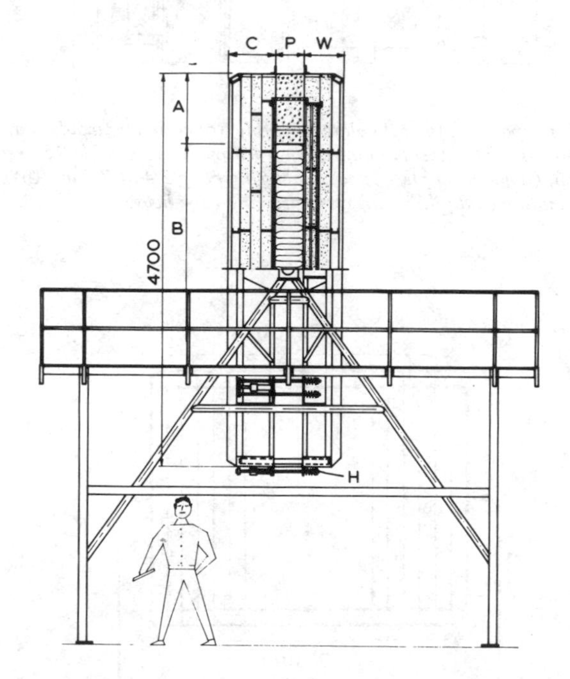

FIG. 3—*General arrangement, dimensions in millimetres.* (A) *convection free perimeter insulation,* (B) *test cavity, size of test specimen 2 by 3 m^2,* (C) *cold plate assembly, temperature of cold plate 0 to −170°C,* (P) *perimeter insulation with test specimen, thickness of a specimen up to 0.5 m, and* (H) *hydraulic jacks with springs.*

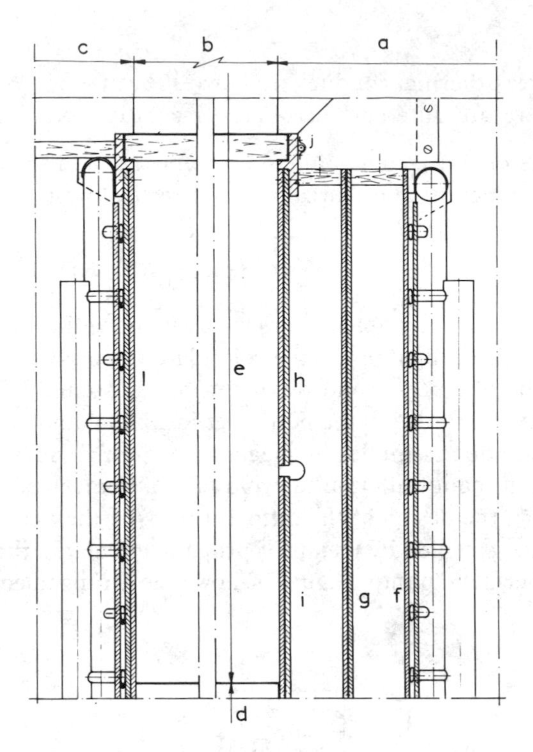

FIG. 4—*Details of apparatus.* (C) *cold plate assembly,* (b) *perimeter insulation, with test specimen,* (a) *warm plate assembly,* (l) *cold plate, with tubes and heating cable,* (d) *test specimen,* (e) *perimeter insulation,* (i) *metering plate, size of metering plate 2.4 by 3.4 m*^{2}, (h) *guard ring,* (g) *guard plate,* (f) *secondary cold plate, and* (j) *guard ring edge heater.*

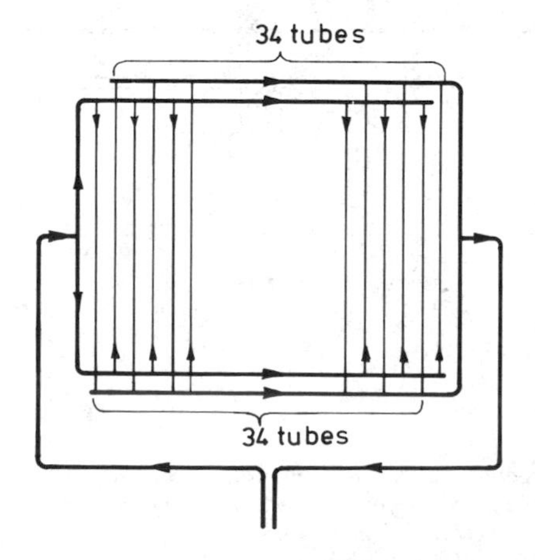

FIG. 5—*Arrangement of tubes in cold plate.*

The refrigerating plant consists of two Stirling cryogenerators in series. The temperature of the cold plate can be regulated between −170 and 0°C by means of the temperature controller acting on the capacity regulation of the cryogenerators. A resistance thermometer in the cold plate is used as a sensor. The cold plate is also provided with heating elements, to be employed in tests where successive cooling down and warming up of the insulating material are to be simulated.

The warm plate assembly, Fig. 6, consists of the main or metering plate, the guard ring, the guard plate, and the secondary cold plate.

When severe convection occurs, the heat load on the lower part of the metering plate will be much larger than the load on the middle and upper parts. To ensure that the warm plate is fairly isothermal also under these conditions, the heating foil of the metering plate is divided into three separate parts, each with its own temperature control. The temperature in the middle of the plate is set by means of a proportional integrator (PI) controller with a resistance thermometer as a sensor, and the temperature of the upper and lower part is regulated to the same level by means of thermopiles and PI controllers. The heating element of the guard ring is also divided into three parts, and controlled by nullamplifiers with relays. The temperature of the guard plate is controlled by means of a thermopile and a PI controller to the same temperature as the metering plate. Thus, by means of the guard ring and the guard plate, it is ensured that all the heat developed in the metering plate passes through the test insulation and the calibrated perimeter insulation.

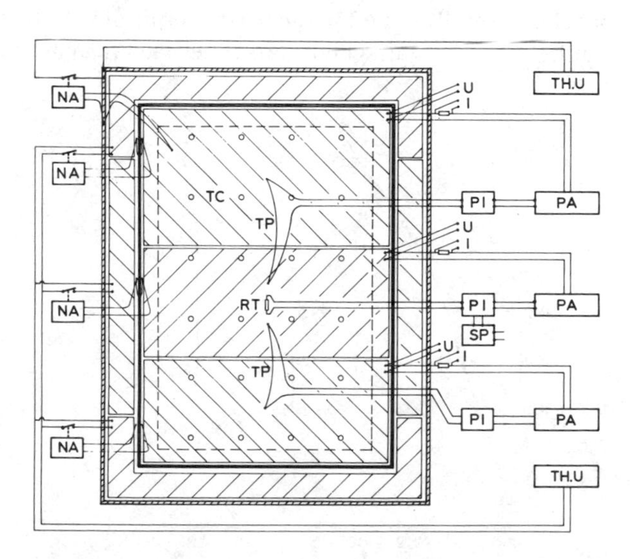

FIG. 6—*Control system for metering plate and guard ring.* (RT) *resistance thermometer,* (TP) *thermopile,* (NA) *nullamplifier with relay,* (TC) *thermocouples,* (U) *voltage measurements, and* (I) *current measurement.*

In order to be able to carry out tests with a warm plate temperature lower than ambient, the secondary cold plate may be cooled by cold alcohol circulated through tubes in the plate. The alcohol is refrigerated by a separate unit, and warm plate temperatures to −30°C may be obtained.

Instrumentation

Temperatures are measured with copper-constantan thermocouples, connected to an automatic data acquisition system. The temperatures of the warm and cold plates are measured with 94 thermocouples embedded in the plates, and in addition the temperature in the test specimen in about 150 points may be measured, dependent on the type of insulation panel to be investigated. The power input to the metering plate is measured by means of the data acquisition system and precision resistors.

Figure 7 shows the temperature distribution on the warm and cold plate, in vertical position, for a specimen with severe convection.

Thermal Stresses

It is evidently difficult to obtain the same thermal stresses in a test insulation as in the actual installation on a spherical tank. Stresses are induced in the test panel in a sufficient extent to disclose defects in design and workmanship, however. This is achieved by ensuring that the thermal contraction of the perimeter insulation follows the thermal contraction of the cold plate, while the test panel is carefully sealed to the perimeter insulation by means of a polyurethane adhesive. To obtain good contact between the perimeter insu-

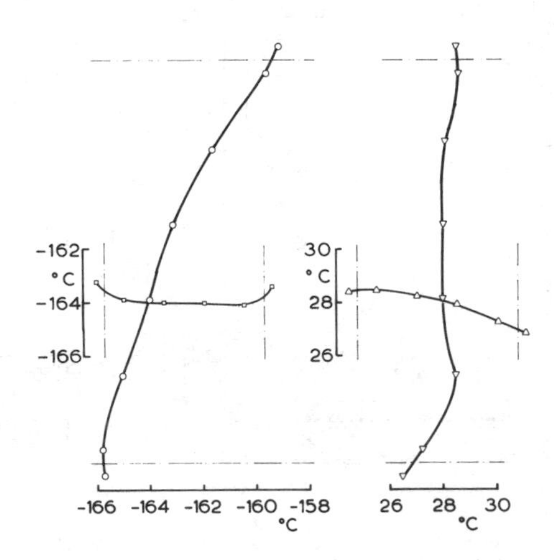

FIG. 7—*Temperature distribution on warm and cold plates, for a specimen with severe convection.*

lation and the cold plate, the glass fiber cloth reinforced perimeter insulation is bonded to a plywood frame, which is resting on the edge of the cold plate. The perimeter insulation is also sealed to the cold plate by means of a compound, giving a firm connection when cooled.

Operating Experiences

The test equipment has been in operation since June 1973. About 30 test panels have been investigated, and 140 test runs have been carried out. The increase in the apparent thermal conductivity due to convection varies with the design of the insulation system, and values up to 200% have been found, Fig. 8 and Table 1. Such inefficient systems were naturally deemed unuseable, and an increase of more than 10% has not been accepted. The large-

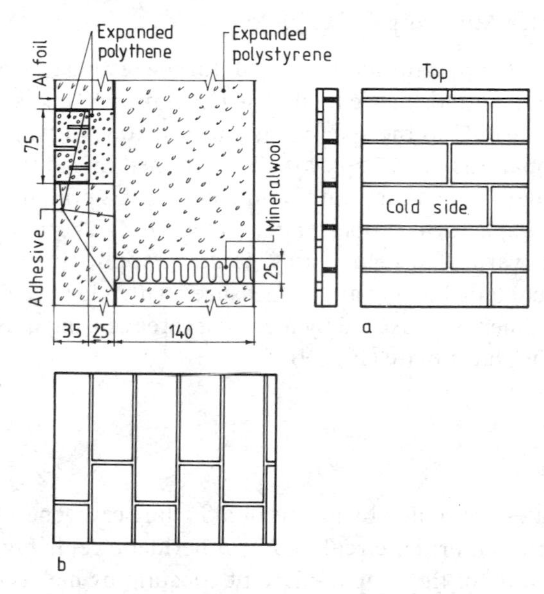

FIG. 8—*Properties of panel with horizontal and vertical expansion strips.* (a) *Vertical and* (b) *90° vertical.*

TABLE 1—*Properties of a panel with horizontal and vertical expansion strips.*

Position	Mean Temperature, °C	Thermal Conductivity, W/(m K)
Horizontal, convection free	−66.7	0.024
Vertical, Fig. 8*a*	−63.5	0.064
Vertical, 90°, Fig. 8*b*	−71.1	0.072

scale guarded hot plate apparatus has proved its practicality as a tool for the evaluation of insulation systems for LNG-tanks.

Testing of Evacuated Insulants

The boil-off from the LNG tank may be reduced by increased insulation thickness, but a thickness above 400 mm may be impractical and space consuming. A reduction in heat conductance may be achieved if the insulation is evacuated. In order to avoid unnecessary weight of the tank system, a thin stainless steel or aluminum membrane is normally used as a vacuum seal. This means that the insulation system must withstand the full atmospheric load.

Heat Flow Meter Apparatus—Design Principle

A hot plate test apparatus has been built for the testing of evacuated insulation systems under atmospheric load. The HFM principle was chosen, since it is easier to apply load to the specimen, and instrumentation is simpler with this type of apparatus. The apparatus is contained in a stainless steel vessel with connection to a vacuum pump, and consists of a primary and secondary cold copper plate, a warm copper plate, and a HFM. Thermocouples are embedded in the warm and cold plates. Load is applied to the specimen by means of a pneumatic bellows and is measured with a load cell. The compression of the specimen is measured by means of three sensing pins, placed at the outer rim of the warm plate (Fig. 9).

Design Details

The cold plates are cooled by means of a Stirling cryogenerator, and liquid propane is used as a brine, circulated by a hermetic centrifugal pump.

The temperature of the propane is kept constant by means of a regulating system, consisting of a heating element, a PI controller and a power source. The warm plate is heated by an electrical element soldered into grooves in the copper plate. The temperature is kept constant by means of platinum resistance thermometers, a PI controller, and a power source.

The HFM is made of quartzfilled epoxy, cast in vacuum, with a copper-constantan thermopile with 60 junctions as a sensing element. The wire diameter is 0.1 mm, and all connections between the two sides of the heat flow meter are made at the outer rim of the plate.

The side towards the test specimen is faced with aluminum sheet, made as three separate rings with radial spacing 1 mm to reduce latent heat flow. The

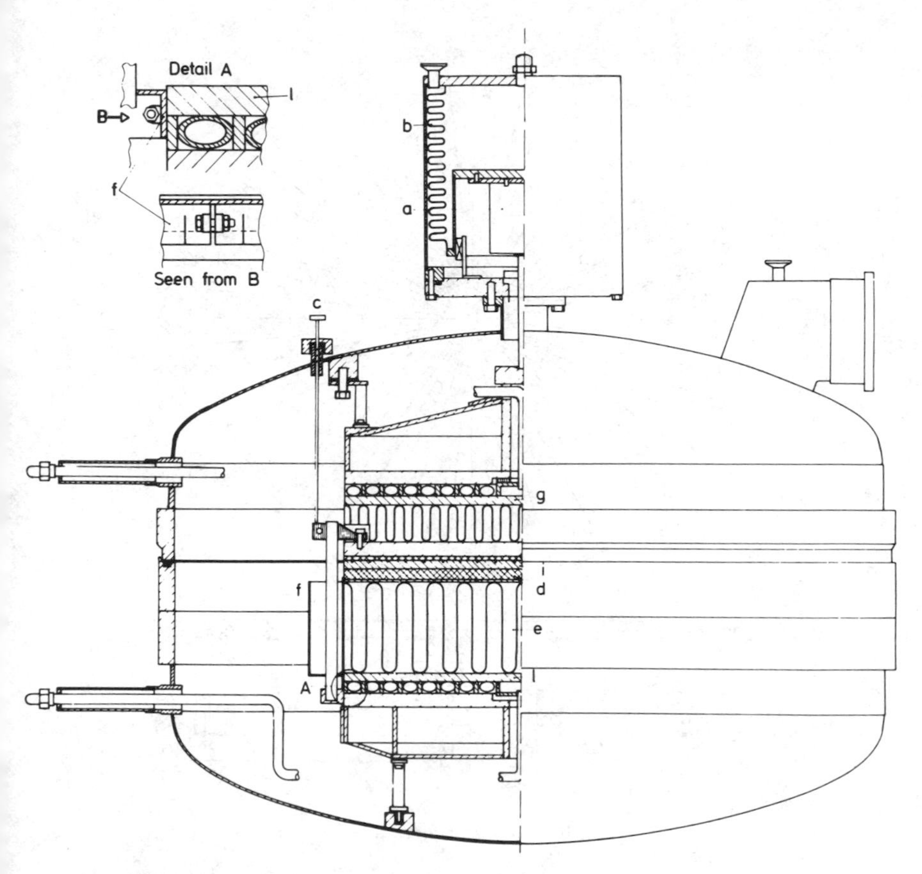

FIG. 9—*Heat flow meter apparatus.* (a) *load cell,* (b) *pneumatic bellows,* (c) *thickness sensing pins,* (d) *heat flow meter,* (e) *test specimen,* (f) *gradient guard,* (g) *secondary cold plate,* (i) *warm plate, and* (l) *cold plate.*

surface temperatures on the warm side are measured by means of thermocouples, embedded in the aluminum sheet. The metering area of the HFM is 100 mm in diameter, with the junctions of the thermopile on a circle with 50 mm diameter, as shown in Fig. 10. A typical calculated temperature distribution is shown in Fig. 11.

The sensitivity of the HFM is approximately 74.2 $\mu V(W/m^2)$.

Calibration is made with high density mineral wool at atmospheric pressure. In order to ensure that no serious inaccuracy occurs when the apparatus is evacuated, the calibration results are controlled by measurements on foam glass in vacuum. The thermal conductivities of the materials used for calibration are measured in a conventional guarded hot plate apparatus.

The vacuum is measured by means of a four-element Pirani gage.

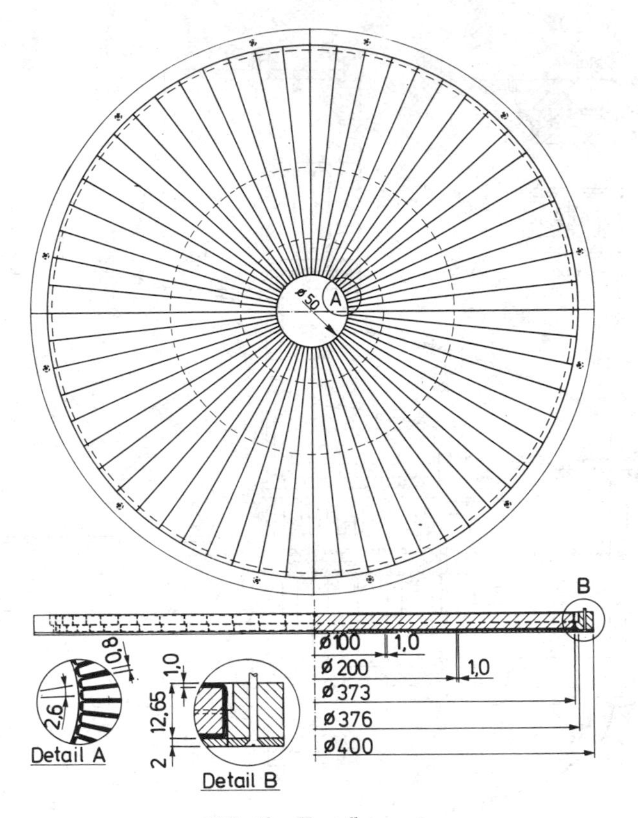

FIG. 10—*Heat flow meter.*

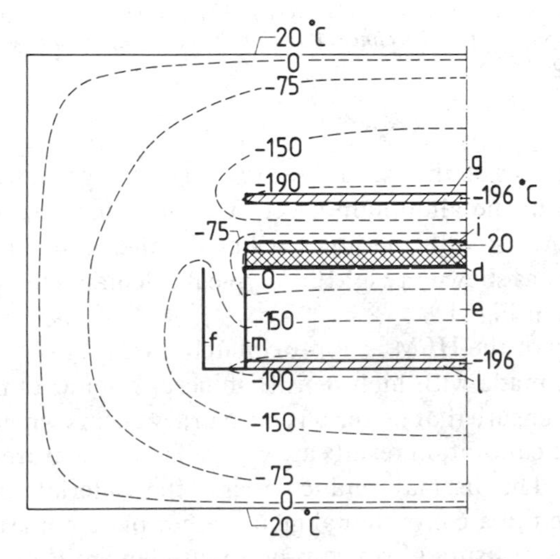

FIG. 11—*Calculated temperature distribution in heat flow meter apparatus. Warm plate temperature 20°C, cold plate temperature —196°C.* (f) *gradient guard, 0.5-mm brass.* (m) *test cavity envelope, 0.8-mm glass fiber reinforced polyester.*

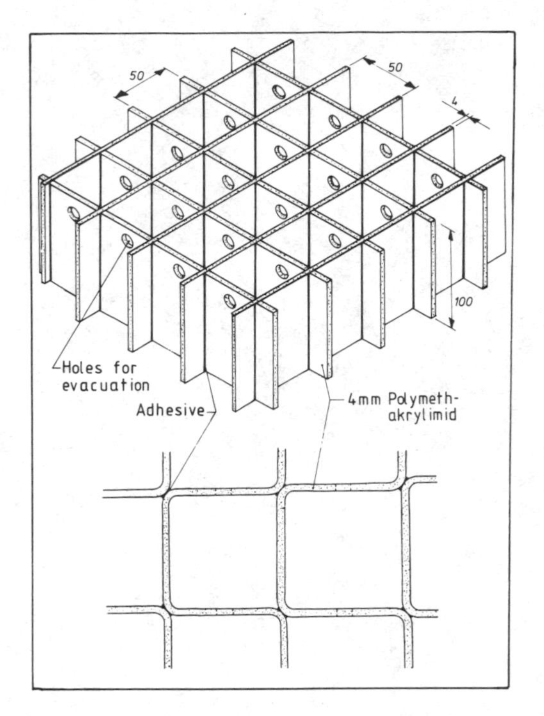

FIG. 12—*Load bearing structure of expanded polymethacylimid for evacuated insulation. In the real application, the material might be heat formed.*

Insulation Systems Tested

Several different insulation systems have been investigated. High-density mineral wool, high-density powder insulants, and honeycomb structures with powder or mineral wool in the cells have been examined. The best system tested so far, however, is a specially built load bearing structure of expanded polymethacrylimid with fibrous glass in the cells, Fig. 12. An apparent thermal conductivity of 0.0057 W/(m K) have been achieved for this system under full atmospheric load, at a pressure in the insulation of 7.10^{-3} mbar. This is naturally not very sensational, compared to an ordinary evacuated insulation, but a reduction of the boil-off from 0.2 to 0.05%/day for a spherical tank of 40 m diameter is within reach.

Apparatus and Instrumentation

Francesco De Ponte[1]

Present and Future Research on Guarded Hot Plate and Heat Flow Meter Apparatus

---

REFERENCE: De Ponte, F., **"Present and Future Research on Guarded Hot Plate and Heat Flow Meter Apparatus,"** *Guarded Hot Plate and Heat Flow Meter Methodology, ASTM STP 879*, C. J. Shirtliffe and R. P. Tye, Eds., American Society for Testing and Materials, Philadelphia, 1985, pp. 101–120.

ABSTRACT: The paper discusses some problems related to the guarded hot plate (GHP) and heat flow meter (HFM) apparatus that justify today research efforts and future research programs. For the GHP these include the problem of imbalance detection and its correlation with the metering area definition, the problem of gap width, and the problem of apparatus speed. Also discussed are the specimen-related issues involved in measurements on rigid, highly conductive materials and mass transfer within the specimen during tests.

For the HFM apparatus the different configurations are surveyed, and the effect of the HFM temperature uniformity and its possible correlation with the HFM thermopile layout are discussed.

In conclusion a concise account is given of the activity within the International Organization for Standardization's Technical Committee 163 on Thermal Insulation in its development of international documents for the two test methods.

KEY WORDS: steady state thermal transmission properties, test methods, guarded hot plate, heat flow meter

When researchers try to investigate unknown aspects of a problem to solve existing doubts, they usually reach, more or less quickly, conclusions that later on form an accepted background in a technical area. Yet, during this path, it often happens that information accumulated up to a certain moment proves only that the real problem is much more complex than the starting one. This could be the case of the guarded hot plate (GHP) apparatus, and to a larger extent, of the heat flow meter (HFM) apparatus.

Some 20 years ago an accuracy of 3 to 5 percentage points was considered

[1]Professor, University of Padova, Istituto di Fisica Tecnica, Padova, Italy.

satisfactory in the majority of thermal conductivity measurements. Such accuracies prevented an exact interpretation of all physical aspects of measured properties, so that errors and differences due to actual physical phenomena were mixed together. Currently an accuracy of 1 percentage point can be claimed by many good laboratories throughout the world. This not only allowed a better understanding of heat transfer mechanisms, but the discovery of its complexity suggested more sophisticated investigations, requiring in turn more accurate measurements. A new interest on the research on accuracy, therefore, was observed in the last years, since the pioneering work made in the 1950s.

The main doubts on the design and use of both the GHP apparatus and the HFM apparatus will now be reviewed that justify today and future research in this field.

Guarded Hot Plate Apparatus

Imbalance Detection, Metering Area Definition, and Gap Width

During the 1950s serious error analysis began. The only possible approach at that time was that of splitting all sources of errors in some simplified configurations allowing not too complex computations. Examples of this analysis are Somers and Cyphers [*1*], Dusinberre [*2*], and Woodside and Wilson [*3–5*]. Most of the results achieved are still valid, yet it will be immediately evident that a GHP design and operation should be based on the consideration of the interaction between the various sources of errors, and it will be evident that problems are not fully clarified.

The cross section of a GHP is shown schematically in Fig. 1; only square GHPs will be considered here; (b is the hot-plate central-section half-width; c and a are the hot plate guard ring internal and external half width; d is the gap width; h is the specimen thickness; ξ is the metal plate thickness, while x_0 to x_9 and z_0 to z_5 are the co-ordinates of the mesh used for a three-dimensional finite-elements GHP analysis summarized here). For the problem of the correct definition of the metering area, it is assumed that the heating power supplied to the central section heater will be uniformly distributed, so that it will cross the specimen through a square area of side $(2b + d + 2\Delta x)$, Δx being the distance of the border of the so-defined metering area from the gap centerline. The widely accepted assumption of the gap centerline as border of the metering area corresponds then to $\Delta x = 0$.

The parameter $\psi = (2\Delta x/d)$ is a measure of the distance of the border of the metering area from the gap centerline, while $\zeta = x/b$ is a measure of the distance from the axes.

T_c is the temperature along the gap of the edge of the central section at a distance ζ from the hot plate axes, and T_g the temperature on the inner side of the guard section in a point along the gap, just in front of the one where T_c is

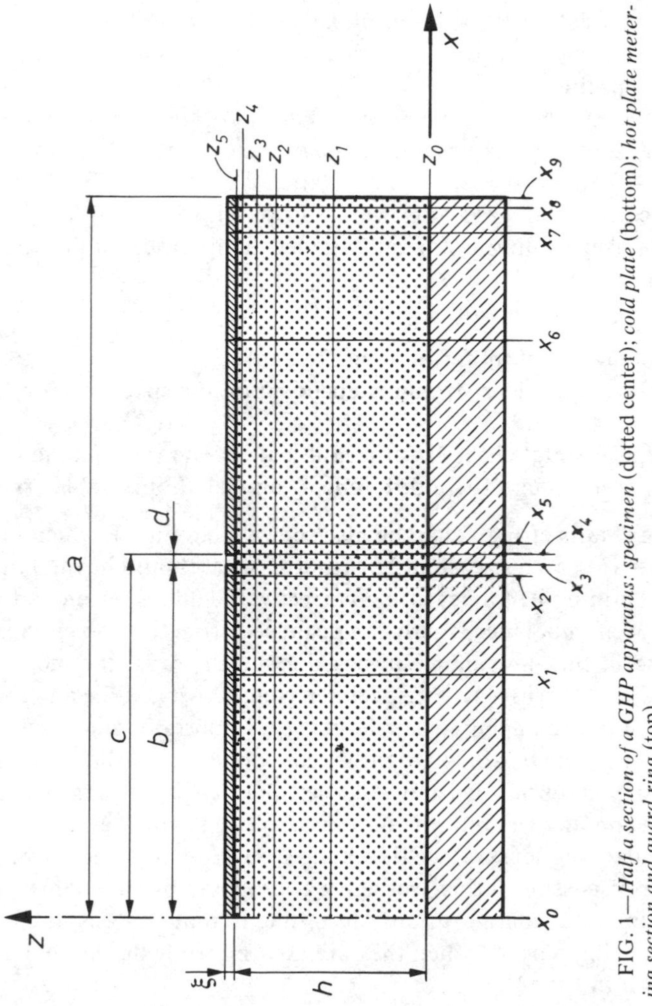

FIG. 1—*Half a section of a GHP apparatus: specimen* (dotted center); *cold plate* (bottom); *hot plate metering section and guard ring* (top).

evaluated. T_o is the temperature of the hot plate center ($x = y = 0$, $z = h$). Assuming the power supplied to the central section heater as the one that, crossing a metering area (defined through ψ) in an ideal indefinite slab bounded by two parallel isothermal planes, would create a temperature difference T_o through the slab, then it is also assumed the power supplied to the guard heater as the one that would create a temperature difference T_o in the center of the GHP apparatus when the central section heater is fed as described. Numerical results derived by De Ponte et al [6] for a set of dimensions a, b, c, d, etc., relative to a particular apparatus are given here just to illustrate the problem.

Figure 2 is a plot in dimensionless form of temperature imbalance $(T_c - T_g)/T_o$ along the gap as a function of the distance ζ from the hot plate axes for $\psi = 0$ and $\psi = -0.2$. Interpolation is possible to find temperature imbalance corresponding to other values of ψ. This is useful to solve the inverse problem of finding ψ for an assigned temperature imbalance and a given distance ζ.

By observing Fig. 2 it can be seen that:

(*a*) imbalance is not uniform along the gap,

(*b*) imbalance profile is strongly dependent on the specimen conductivity, and

(*c*) an experimental check is rather difficult due to the small temperature differences, even though these differences can yield appreciable errors.

The same results of Fig. 2 can be also shown as in Fig. 3, which illustrates what the metering area size should be (that is, what should be the value of the ordinate ψ defining it) to measure a correct value on a specimen of thermal conductivity λ if imbalance is zero at a distance ζ from the axes. It can be seen that for $\zeta = 0$, that is, balance imposed along the axes, the metering area definition is very sensitive to the specimen conductivity, while an error in the positioning of the balancing sensors is not very important (curves corresponding to ζ ranging from 0 to 0.34 are rather close, one to the other). On the other hand a balancing sensor placed in a corner, that is, $\zeta = 1$, is far less sensitive to specimen conductivity but much more sensitive to small errors in the positioning of balancing sensors (see the changes on ψ when ζ ranges from 0.68 to 1). A deeper discussion of this problem was proposed by De Ponte et al [6] for a limited number of configurations; according to that analysis, the best position for balancing sensors, when they are not uniformly distributed along the gap, is shown in Fig. 4.

Figure 3 points up that, for the apparatus considered, there was an uncertainty of the order of ± 5 percentage points in defining ψ. This consideration would show that the adoption of small gaps is required to minimize the effect of the uncertainty in the definition of the metering area. It would be also possible to reduce this uncertainty by designing and operating the apparatus in a narrow range of conductivities or by selecting suitable positions for the bal-

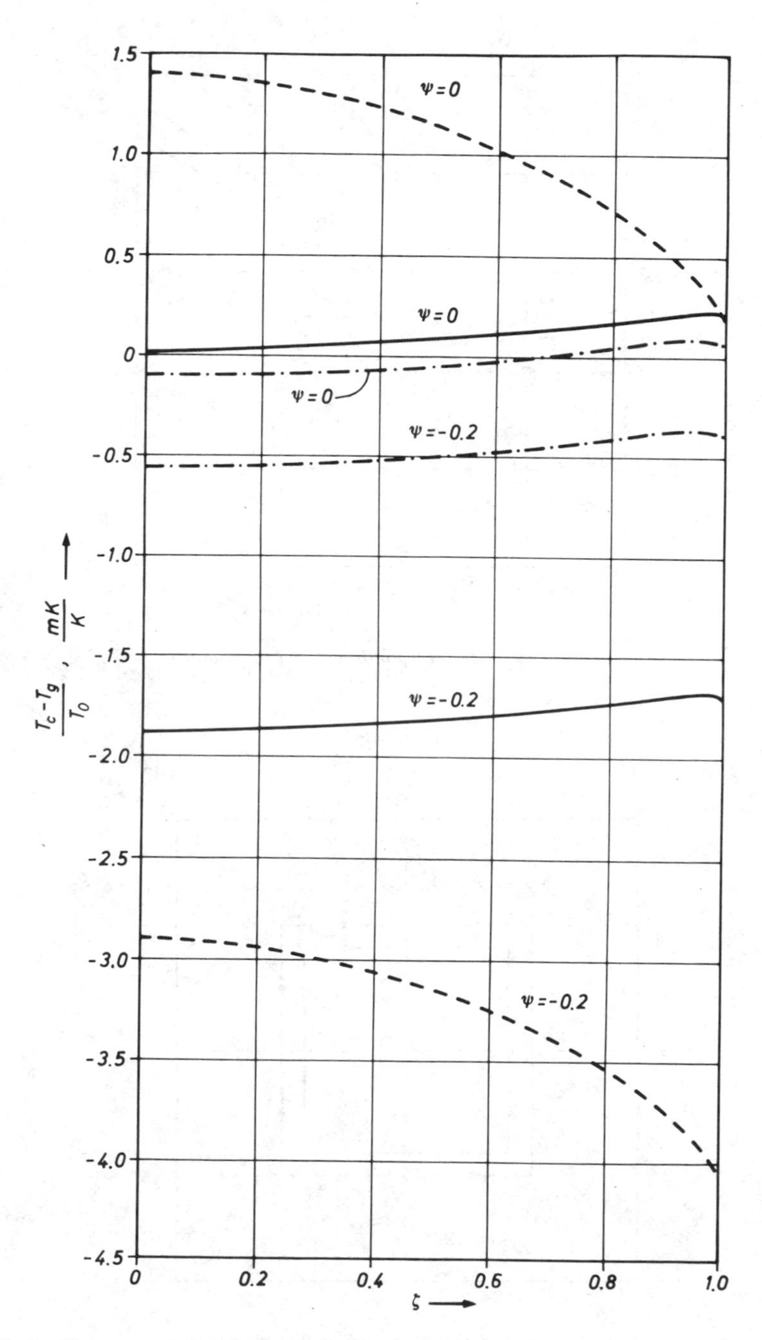

FIG. 2—*Temperature imbalance for* $\lambda = 0.01\ W\ m^{-1}\ K^{-1}$ (dashed and dotted lines); *for* $\lambda = 0.1\ W\ m^{-1}\ K^{-1}$ (continuous lines) *and for* $\lambda = 1\ W\ m^{-1}\ K^{-1}$ (dashed lines).

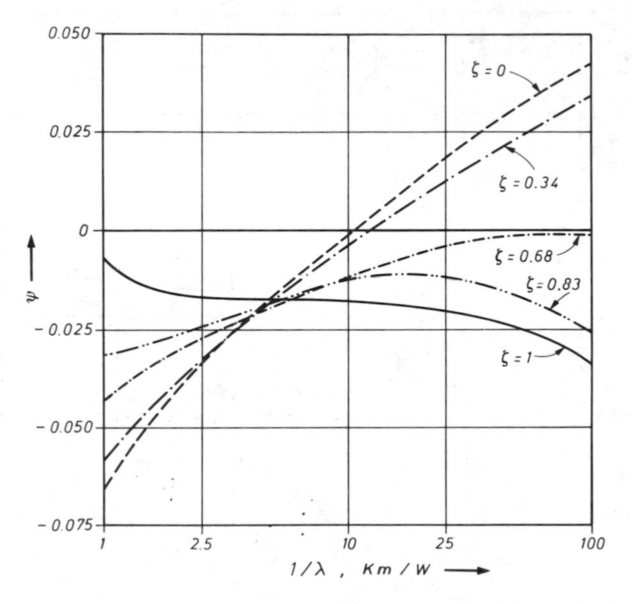

FIG. 3—*Values of ψ corresponding to zero imbalance.*

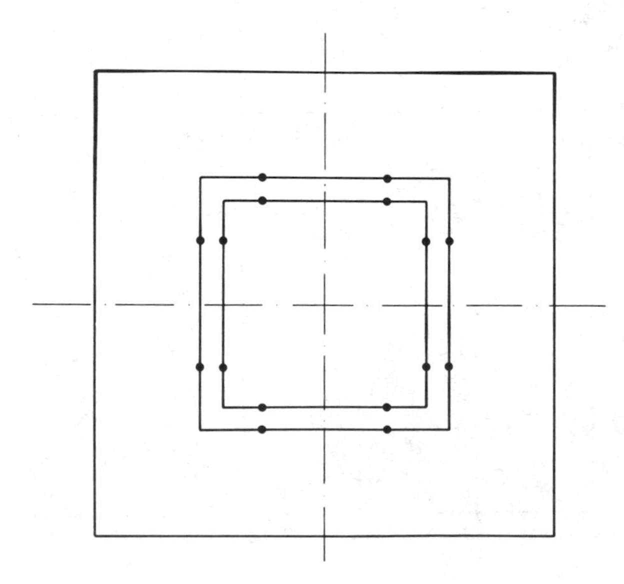

FIG. 4—*Suggested positions for balancing thermocouples.*

ancing sensors, anyhow there is no doubt that the first reaction to Fig. 3 remains that of reducing the gap width.

Similar conclusions are reached by the consideration of Fig. 5, dealing with the problem of testing very thin specimens. The flow lines show that the metering area definition coincides with the edge of the hot plate central section when the specimen thickness tends to zero. Again, in order to avoid complex corrections due to specimen thickness, the gap width should be far smaller than the minimum specimen thickness to be tested.

Unfortunately, as frequently happens, only one side of the whole problem has been considered. A small gap width means low guard-to-center resistance, that is, low imbalance sensitivity and thus a good probability of observing large imbalance errors. There is an optimum gap width that will minimize the sum of all errors. Unfortunately, while the uncertainties in the definition of the metering area span from the gap centerline to the gap edge, that is, the maximum theoretical error is known exactly, the imbalance error is a complex function not only of the gap and specimen geometry and of the specimen thermal resistance, but also of the position and layout of the balancing sensors.

Figure 6 is a sketch of some existing solutions, while Fig. 7 can help in understanding the influence of uneven distributions of contact resistances between sensing elements and metal plates or, in some cases, between sensing elements and the heater or the specimen. Figure 7 also illustrates the importance of having large resistances, as compared with the previously mentioned contact resistances, between the sensors placed on the opposite sides of the gap.

Another issue frequently disregarded is the distance of the imbalance sensing elements from the gap edge. Fig. 8 is an illustration of a temperature profile of the surface temperatures in a cross section of a specimen. The curved temperature distribution on the surface in contact with the guard ring is due to edge heat losses both through the specimen edges and through the guard ring edges and the electrical and mechanical connections of the hot plate to the rest of the apparatus. It is quite evident that imbalance sensors placed in a couple of points such as A and B in Fig. 8, that is, far away from

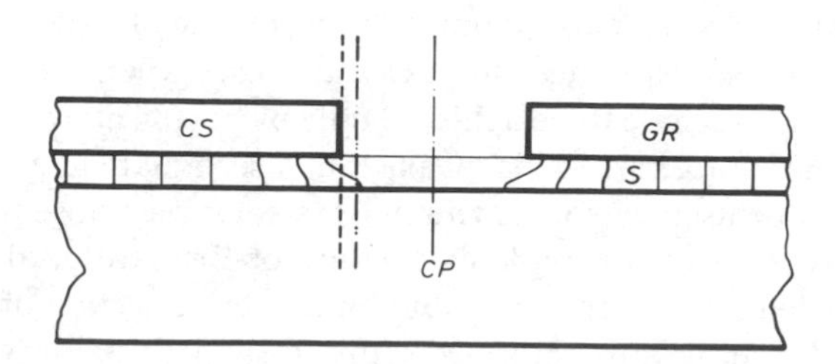

FIG. 5—*A thin specimen in a GHP apparatus. CS: hot plate central section; GR: hot plate guard ring; CP: cold plate; S: specimen.*

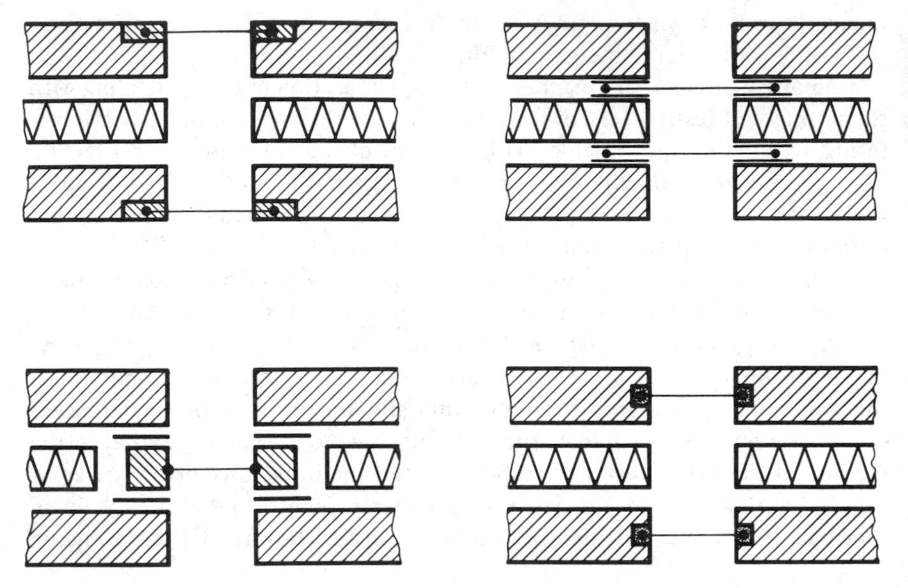

FIG. 6—*Some possible arrangements of balancing sensors.*

the gap, will force, when balanced, a temperature difference just through the gap, creating an imbalance heat flow and consequently a measurement error. It is therefore selfevident that the surface sensors that detect the temperature uniformity all over the hot plate central section and hot plate guard ring cannot be used to detect imbalance too.

All of the considerations require a knowledge of the temperature distribution in the whole hot plate. Unfortunately the majority of GHP modelling for error analysis assumed hot plate central section, hot plate guard ring, and cold plates at uniform temperatures, and did not take into account the effect of temperature profiles on the apparatus surfaces. Donaldson [*7*] tried a solution in 1961, while more recently comprehensive models were developed to support the drafting of a standard on the GHP method within the International Organization for Standardization (ISO); Bode faced the problem through series expansions and Troussart [*8*] through a finite-element analysis of some apparatus. Nevertheless, from the report of Ziebland [*9*] on an interlaboratory comparison among four laboratories, experimental evidence shows that best accuracies can be achieved only by paying attention to the hot plate temperature uniformity. This means that a secondary guard or a gradient guard should be used whenever running tests not very close to room temperature. Figure 9 shows a sample calculation of Bigolaro et al [*10*] just to verify the improved temperature distribution in a cross section of a specimen mounted in a GHP apparatus equipped with a *T*-shaped auxiliary guard. The improved temperature distribution in the specimen should thus result in an improved temperature uniformity in the hot plate.

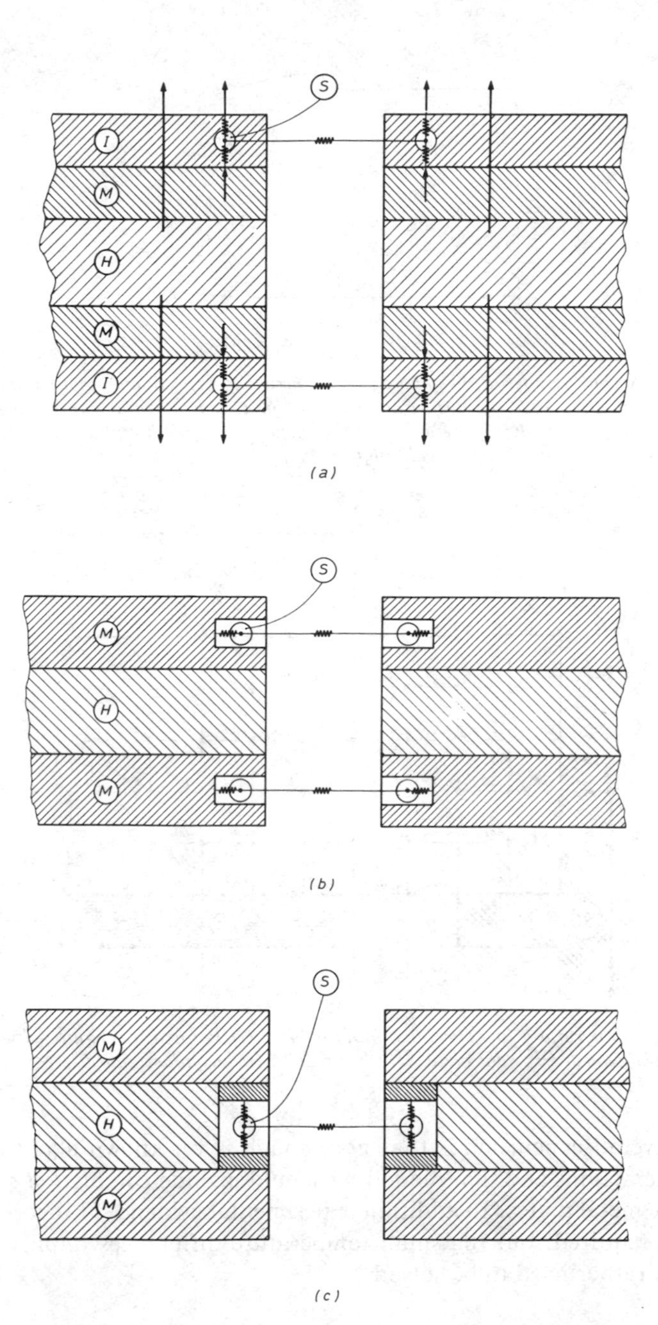

FIG. 7—*Thermal resistances in imbalance analysis. S: imbalance sensor; M: central section or guard ring metal plate; H: central section or guard ring heater; I: electrically insulating sheet with embedded imbalance sensors.*

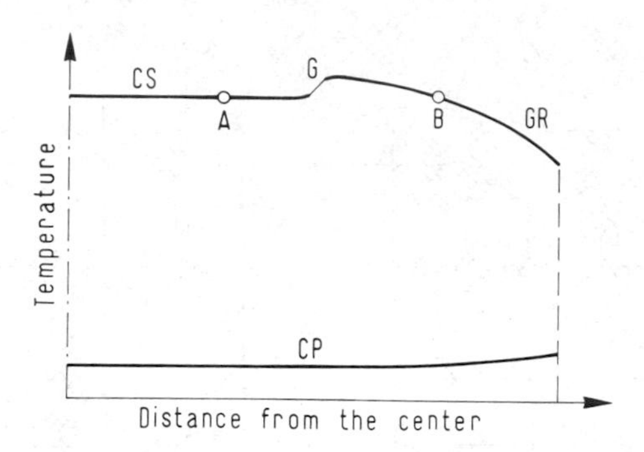

FIG. 8—*Surface temperatures in a half cross section of a specimen mounted in a GHP apparatus. CS: temperatures of the surface along the hot plate central section; G: temperatures of the surface along the gap; GR: temperatures of the surface along the hot plate guard ring; CP: temperatures of the surface along the cold plate.*

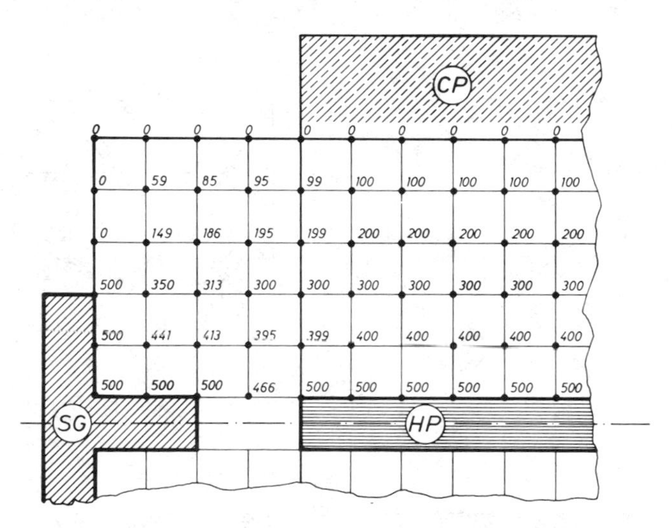

FIG. 9—*Temperatures in a specimen mounted in a GHP apparatus equipped with a secondary* T-*shaped guard. CP: cold plate; HP: hot plate; SG: secondary guard.*

The overall conclusion is that not enough is yet known about errors and their interactions and that modelling should include conduction through the hot plate metal plates. In addition experimental evidence of theoretical modelling is required, but the small temperature differences implied make the problem rather hard to be solved.

Measurement Speed

For rapid apparatus operation automatic controllers are requested not only to balance guard ring and central section, but also to impose the central section temperature. The frequent practice of feeding the central section heater through a stabilized d-c supply does not meet this requirement: this solution fixes just the heat transfer rate, not the temperature.

The problem connected with the automatic control of the central section temperature is that both temperature and electrical power shall be measured during steady state to compute thermal properties. To obtain not only a constant temperature but also a constant power, the controller shall not be compelled to react to environment disturbances. This can be achieved only by the elimination of disturbances, that is, by enclosing the apparatus in a carefully conditioned cabinet.

The controller itself must not introduce electrical noise: under this respect a very low-noise linear amplifier is the best choice of controller (that is, just the proportional action and no reset or derivative action is recommended for the controller). Bigolaro et al [*11*] proved that this solution is satisfactory, but a systematic study on automatic control of GHP apparatus is required, considering also the new solutions offered by the use of microprocessors and digital controllers.

The Specimen

The specimen type and form also raise many unsolved doubts that justify further research. Some of these are due to the fact that most error analysis is based on homogeneous opaque isotropic specimens, while many of the materials and systems to be evaluated are:

(*a*) nonisotropic such as fibrous or composite insulations,
(*b*) multi-layer such as reflective insulations,
(*c*) metal-foil faced such as polyurethane boards, and
(*d*) semi-transparent such as light-density insulations.

How do these affect estimates of imbalance and edge heat losses?

A considerable amount of experimental work has been conducted on light density insulations to prove that their behavior diverges from that of an homogeneous specimen. As examples, opaque septa were used by Pelanne [*12*] to show the presence of radiation and to separate errors from physical phenomena, while other researchers covered specimen edges with reflective foils to reduce edge heat losses due to radiation heat transfer. In addition a great deal has been written on heat transfer by radiation in light-density insulants to explain the overall heat-transfer mechanism, thus allowing us to give also an explanation of the so-called thickness effect. Examples of this work include, for example, Linford et al [*13*], Larkin and Churchill [*14*], and Rennex [*15*]. However, little has been done to predict errors (edge heat losses first of all) when radiation heat transfer and radiation scattering in particular are a relevant part of the total heat transfer through the specimen. This was possibly a source of disagreement in the evaluation of the thickness effect.

Considering specimens of hard, highly conductive materials like concrete, a laboratory procedure sometimes followed is that of splitting the specimen in two sections: one of the same size as the central section and the other of the same size as the guard ring. A gap therefore exists also in the specimen, as shown in Fig. 10. As the gap space creates a discontinuity in the specimen, this gap is usually filled with a low-conductivity material. Bode has developed some computations for the aforementioned ISO draft standard on the GHP method, but just a few researchers have tried to evaluate metering area and imbalance errors in this complex configuration, so that no universally accepted reply exists on the limits of applicability of such a procedure.

A fully different problem related to the specimen is its amount of moisture content and the effect of mass transfer during the measurement of steady-state thermal transmission properties. It is well known that moisture effect must be carefully considered in any measurement on inorganic highly conductive materials like concrete. Many researchers, for example, Tye and Spinney [*16*] have shown that differences in moisture contents in these materials can result in large changes in thermal properties. Yet the effect of mass transfer must be carefully investigated even in any measurement on thermal insulations whenever high accuracy (for example, better than one percentage point) is required.

In dealing with moisture, the term "dry state" should be defined. A universal definition is not available, but, whatever the definition of the conventional dry state, tested specimens are never perfectly dry, so that the effect of moisture redistribution within the specimen during the test execution can be seen. Figures 11 and 12 illustrate measurements on high-density resin-bonded glass-fiber boards: Fig. 11 shows the effect of changing test temperature from 136 to 165 K. As testing temperatures are both very low, free moisture is quite absent, thus allowing rapid evaluations. Tested data are again within 1 percentage point of steady-state values in 3 h.

The lower plot of Fig. 12 shows a change in testing conditions closer to room temperature on the same specimen and same apparatus as in Fig. 11.

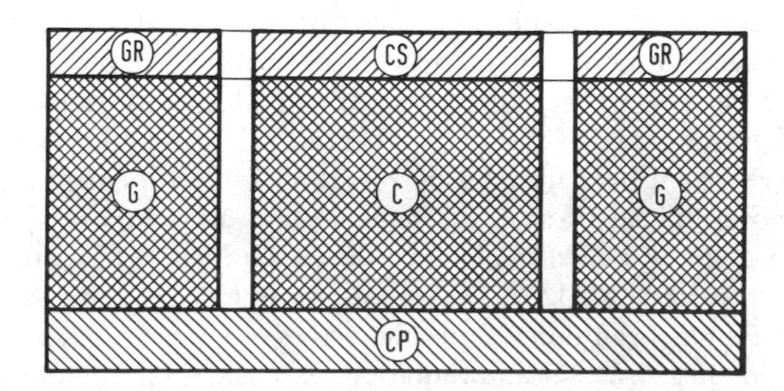

FIG. 10—*Splitted specimen in a GHP apparatus. CS: hot plate central section; GR: hot plate guard ring; CP: cold plate; C: specimen central section; G: specimen guard ring.*

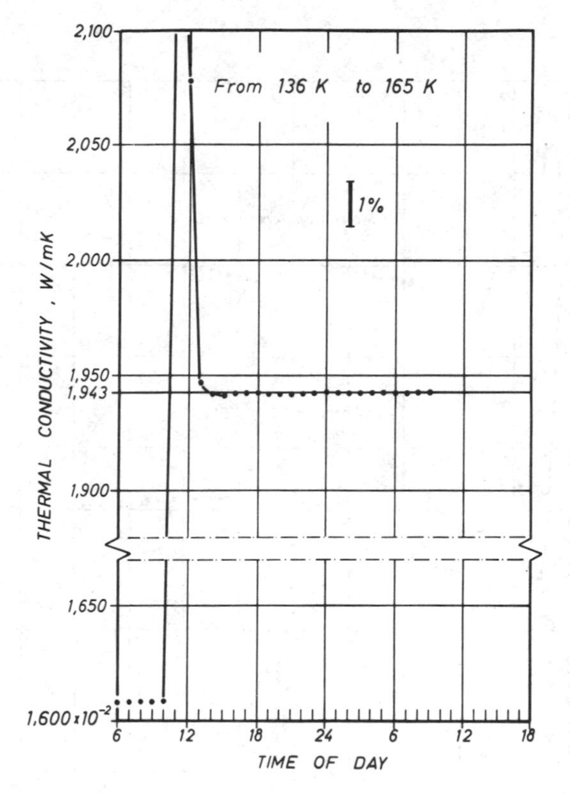

FIG. 11—*Measured thermal conductivity of a high-density fibrous specimen during a transient phase at low temperatures.*

Here mass transfer within the specimen overlaps heat transfer, but it is very much slower.

The upper plot of Fig. 12 shows then what happened to the same specimen tested in the same apparatus when just the relative humidity of the air of the cabinet enclosing the apparatus was modified by injecting small amounts of water vapor. Further details are given by Bertasi et al [*17*].

Heat Flow Meter Apparatus

Three possible configurations are known for the HFM apparatus, as shown in Fig. 13. The solution with a single specimen and a single heat flow transducer (HFT) as in Fig. 13*a* is the simplest, while the most complex is that of Fig. 13*b* with two HFTs. A more comprehensive discussion on the comparison of the three solutions and on some HFM problems was proposed by De Ponte and Maccato [*18*], and the major conclusions are discussed next. Figures 14*a* and 14*b* supply the theoretical ground to evaluate the speed in

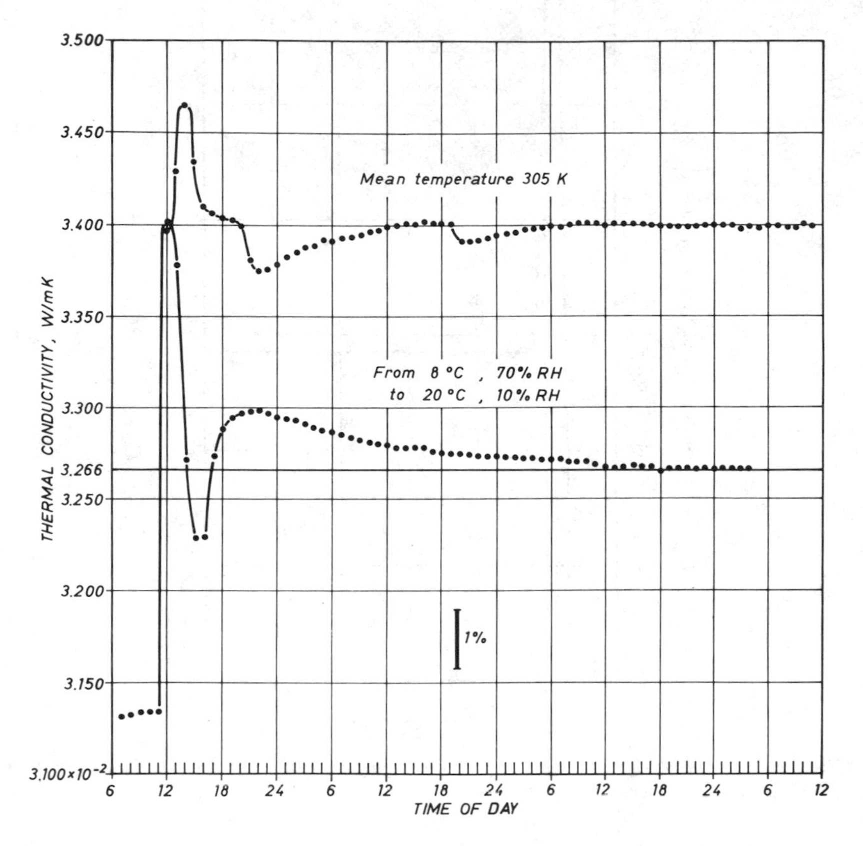

FIG. 12—*Measured thermal conductivity of a high-density fibrous specimen during a transient phase near to room temperature* (lower plot) *and while changing relative humidity* (upper plot).

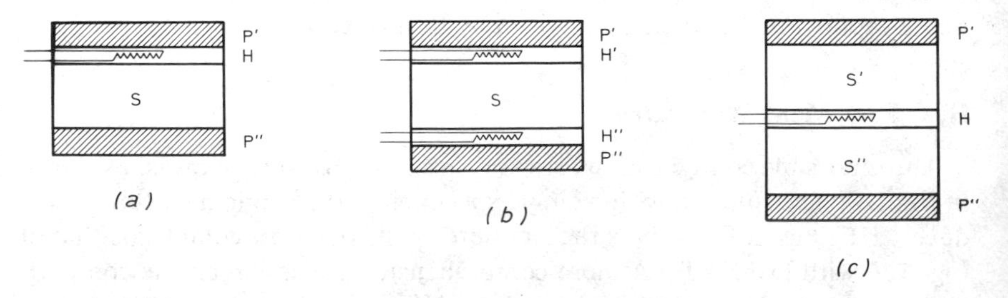

FIG. 13—*Some configurations of HFM apparatus. H: heat flow transducer; P: heating or cooling plate; S: specimen.*

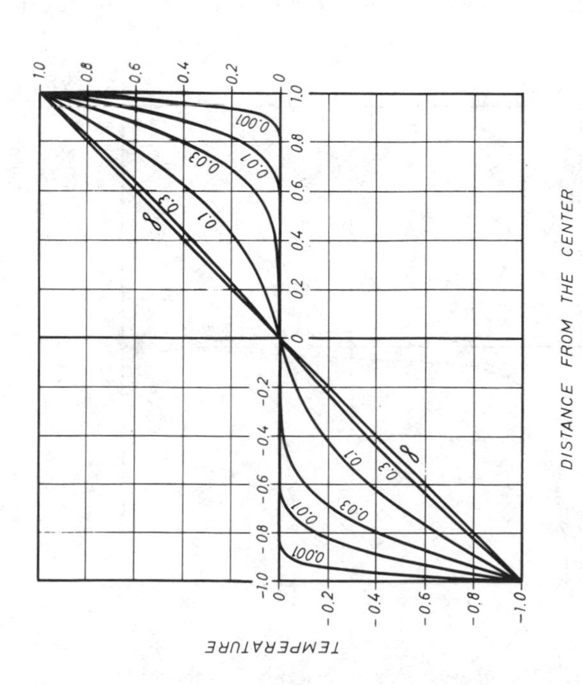

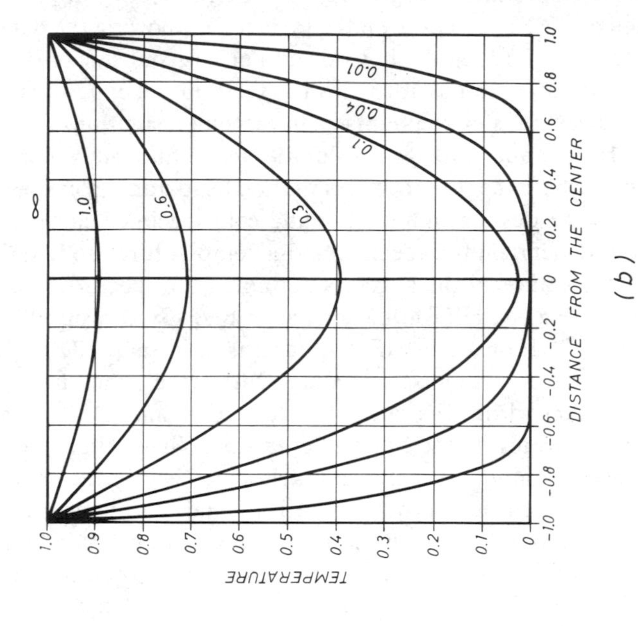

FIG. 14—*Temperature distribution in a slab of thickness* d = 2, *initially at temperature* t = 0 *and then* (a) *with one side at temperature* t = +1, *the opposite at temperature* −t; *and* (b) *with both sides kept at temperature* t = 1. *The parameter of the curves is the Fourier number* $4a\tau/d^2$, a *being the thermal diffusivity and* τ *the time.*

reaching the steady state with the different HFM configurations. It is shown that the one with one specimen and two HFTs is more speedy than that with one specimen and one HFT. In addition the former, by allowing the detection of the heat-transfer rate both on the hot side and on the cold side of the same specimen, opens frontiers not possible with the GHP in the experimental investigation on the heat transfer mechanisms in thermal insulations. Carefully calibrated two HFT apparatus are valuable tools for analyzing semi-transparent media, where the presence of radiation heat-transfer can result in unequal heat-transfer rates through the hot and cold specimen side (unequal heat-transfer rates depend on the specimen-edge temperatures and on the total emispherical emittances of the surfaces bounding the specimen edges).

Another field where a two HFT apparatus as in Fig. 13*b* can supply better information than a GHP or a one HFT apparatus as in Fig. 13*a* is that of measurements on more or less moist materials. During these measurements a moisture redistribution within the specimen can exist that results again in unequal heat transfer rates on the hot and cold side. These phenomena are usually much slower than heat transfer in reaching a steady-state condition. Thus slow changes in measured properties can be masked by measurement inaccuracies on a GHP or a HFM apparatus as in Fig. 13*a*, but steady state is surely not achieved as long as heat transfer rates are not equal on both the hot- and the cold-side HFT. Figure 15 shows a plot of test data in such a situation: it should be pointed up that the reading of both HFTs is not only an easy means of detecting the attainment of the steady state, but that the aver-

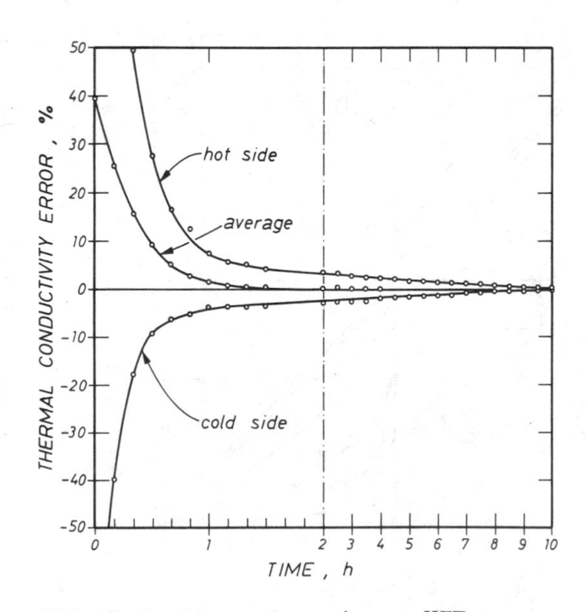

FIG. 15—*Reaching steady state in a two HFT apparatus.*

age of such readings approximates the steady-state value much earlier than each HFT reading.

It has been pointed up that the HFM apparatus, even though simpler than the GHP apparatus, can supply information not given by the GHP. In addition it can reach reproducibilities equal or better than those of the GHP and can reach accuracies only slightly worse than those of the GHP. Nevertheless the HFM apparatus is far less known and researched than the GHP.

Little is known about its modelling or its error analysis. Just as an example, Fig. 16 is a plot of the influence of the environment temperature on measured data (edge heat loss error) for an apparatus where the insulation of the heating and cooling plates was increased on the side not facing the specimen. The first consequence was an increased temperature uniformity for the hot and cold plates, due to the reduced heat transfer towards the environment, but this resulted in a modified sensitivity of the HFTs to edge heat losses (note that hot side edge heat loss error changed even in sign). A possible explana-

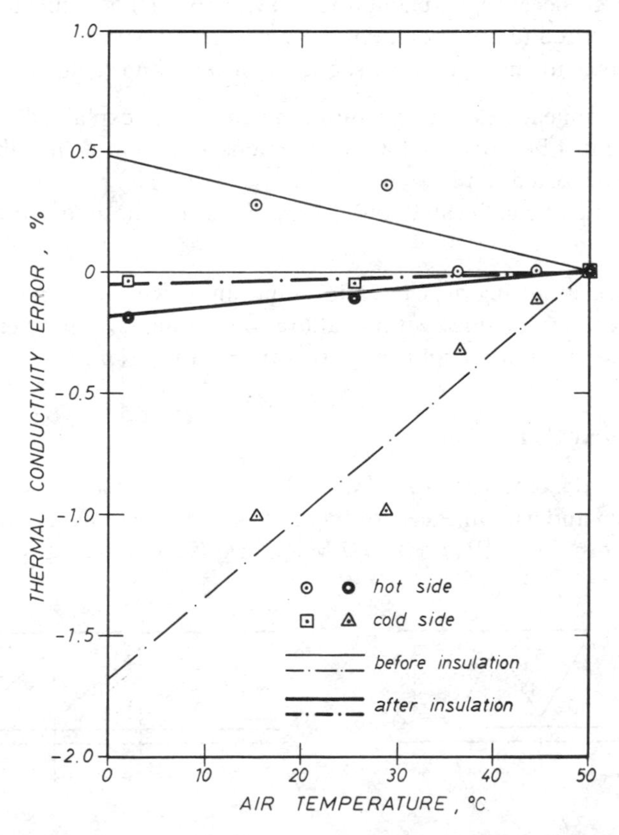

FIG. 16—*Influence of surrounding air temperature on measured thermal conductivity in a two HFT apparatus.*

tion was sought considering the HFT thermopile layout; from Fig. 17 it is evident that thermopile layout (*a*) is sensitive to temperature gradients along the main dimensions of the HFT, while layout (*b*) is not.

The sensitivity of the HFM itself to edge heat losses is therefore totally different from that of the GHP. Even more complex effects of edge heat losses are expected with nonisotropic, semi-transparent, or layered specimens. It is possible that new criteria are needed for the heating and cooling plate temperature uniformity, and there is not yet evidence whether today thinner and faster HFTs are more or less sensitive to a plate temperature nonuniformity.

From these short considerations it is possible to apply GHP error analysis also to measurements with HFMs only when heat transfer within the specimen is considered, as in the case of edge heat losses just through the specimen. However, totally different approaches are required in order to consider the errors connected with heat transfer through the heating and cooling plates. (See the previous discussion for the HFM apparatus, and the imbalance problem of Fig. 8 for the GHP apparatus.)

In conclusion, because not enough is known about HFMs, further investigations are required to predict and maintain all errors within narrow and acceptable limits. Currently, as a consequence of poor knowledge:

(*a*) some nonlinearities during calibration cannot be explained,
(*b*) HFMs must be calibrated with specimens as similar as possible to the specimens to be tested, and
(*c*) HFMs must be calibrated under conditions very close to actual testing conditions.

This is because, during present calibration, unknown errors that in reality are not constant are included within calibration constants. The research goal should be therefore that of splitting errors from calibration.

ISO Activity and Conclusions

ISO Technical Committee 163 (TC 163) on Thermal Insulation started its activity in 1976 and was an excellent forum to share experiences and possible solutions on both the GHP and the HFM apparatus. In some cases it was also

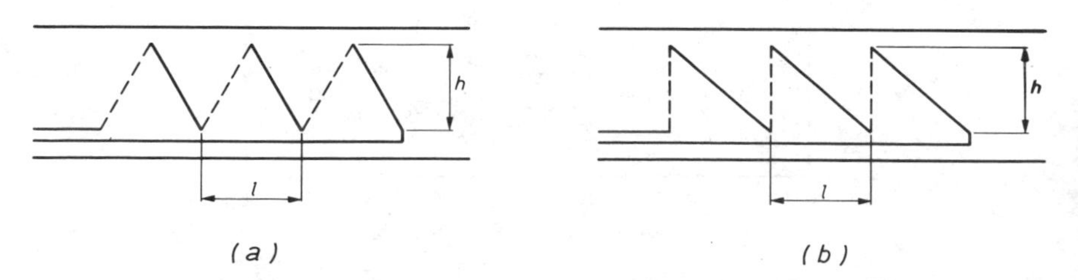

FIG. 17—*Thermopile-junction layouts in a HFT.*

an opportunity to develop new high-level research, as the already recalled works of Bode and Troussart. Efforts were made to collect all known solutions while pointing up their limits of applicability. This work ended in two very comprehensive documents, one on the GHP apparatus and the other on the HFM apparatus. The major conclusions, starting with the GHP apparatus can be summarized as follows:

1. Accuracy better than one percentage point can be achieved around room temperature.
2. A second guard or a gradient guard is mandatory except for tests at room temperature.
3. Only few balancing solutions are satisfactory when considering the balancing problem under all possible implications.
4. Many tools exist to predict GHP apparatus accuracy during its design process.
5. Most of the apparatus performances can be also experimentally checked.
6. The concept of steady state is rather complex, and many national standards fail on this point.
7. Automatic control of both central section and guard ring heaters speeds up measurements, thus reducing speed limits of GHP versus HFM apparatuses.

Within ISO TC 163, Subcommittee 1, separate working groups were charged to the drafting of documents on the GHP and HFM methods, respectively. The membership of the two working groups were mostly different, but very good cooperation was established and maintained, yielding an approach to the two documents which was the same.

During the development of the ISO document on the HFM it was evident that HFMs are cheaper and simpler in use than GHPs but:

1. Many national standards on HFT are simpler and shorter than the corresponding GHP standards only because less is known about the HFM: a deeper error analysis is required.
2. Temperature uniformity requirements shall be faced in a completely different way for the GHP and the HFM (they could be even more stringent for the HFM and should be related to the HFM structure).
3. The present lack of information can be improved upon only through a wider set of reference materials becoming available.

References

[1] Somers, E. V. and Cyphers, J. A., *Review of Scientific Instruments*, Vol. 22, 1951, pp. 583–585.
[2] Dusinberre, G. M., *Review of Scientific Instruments*, Vol. 23, 1952, pp. 649–650.
[3] Woodside, W. in *Symposium on Thermal Conductivity Measurements and Applications of*

Thermal Insulations, ASTM STP 217, American Society for Testing and Materials, Philadelphia, 1957, pp. 49-64.

[*4*] Woodside, W. and Wilson, A. G. in *Symposium on Thermal Conductivity Measurements and Applications of Thermal Insulations, ASTM STP 217*, American Society for Testing and Materials, Philadelphia, 1957, pp. 32-48.

[*5*] Woodside, W., *Review of Scientific Instruments*, Vol. 28, 1957, pp. 1033-1037.

[*6*] De Ponte, F., Mariotti, M., and Strada, M. in *Proceedings*, XVth International Congress of Refrigeration, Venice, 1979, Vol. II, pp. 653-662.

[*7*] Donaldson, I. G., *Quarterly of Applied Mathematics*, Vol. 19, 1961, pp. 205-219.

[*8*] Troussart, L. R., *Journal of Thermal Insulation*, Vol. 4, 1981, pp. 225-254.

[*9*] Ziebland, H., "Certification Report on a Reference Material for the Thermal Conductivity of Insulating Materials between 170 and 370 K. Resin-bonded Glass Fibre Board. (BCR RM 65)," Bureau Communautaire de Référence, Bruxelles, 1981.

[*10*] Bigolaro, G., De Ponte, F., and Fornasieri, E., in *Proceedings*, XIV International Congress of Refrigeration, Moscow, 1975, Vol. II, pp. 435-442.

[*11*] Bigolaro, G. and De Ponte, F. in *Proceedings*, Meeting Commission B1 of the International Institute of Refrigeration, Washington, DC, 1976, pp. 35-47.

[*12*] Pelanne, C. M. in *Thermal Insulation Performance, ASTM STP 718*, American Society for Testing and Materials, Philadelphia, 1980, pp. 322-334.

[*13*] Linford, R. M. F., Schmitt, R. J. and Hughes, T. A. in *Heat Transmission Measurements in Thermal Insulations, ASTM STP 544*, American Society for Testing and Materials, Philadelphia, 1974, pp. 68-84.

[*14*] Larkin, B. K. and Churchill, S. W., *Journal*, American Institute of Chemical Engineers, Vol. 5, No. 4, 1959, pp. 467-474.

[*15*] Rennex, B. G., *Journal of Thermal Insulation*, Vol. 3, 1979, pp. 37-61.

[*16*] Tye, R. P. and Spinney, S. C. in *Proceedings*, Meeting Commission B1 of the International Institute of Refrigeration, Washington, DC, 1976, pp. 119-127.

[*17*] Bertasi, M., Bigolaro, G., and De Ponte, F. in *Thermal Transmission Measurements of Insulation, ASTM STP 660*, American Society for Testing and Materials, Philadelphia, 1978, pp. 30-49.

[*18*] De Ponte, F. and Maccato, W. in *Thermal Insulation Performance, ASTM STP 718*, American Society for Testing and Materials, Philadelphia, 1980, pp. 237-254.

David L. McElroy,[1] *Ron S. Graves,*[1] *David W. Yarbrough,*[1] *and J. Peyton Moore*[1]

A Flat Insulation Tester That Uses an Unguarded Nichrome Screen Wire Heater

REFERENCE: McElroy, D. L., Graves, R. S., Yarbrough, D. W., and Moore, J. P., **"A Flat Insulation Tester That Uses an Unguarded Nichrome Screen Wire Heater,"** *Guarded Hot Plate and Heat Flow Meter Methodology, ASTM STP 879*, C. J. Shirtliffe and R. P. Tye, Eds., American Society for Testing and Materials, Philadelphia, 1985, pp. 121-139.

ABSTRACT: An unguarded, unidirectional heat flow technique is described for measuring the apparent thermal conductivity (λ_a) of insulating materials in the range 300 to 330 K. Vertical heat flow is generated by a 0.9 by 1.5 m electrically heated nichrome screen that is instrumented and sandwiched between two specimens. The specimens are bounded by two temperature-controlled copper plates. The system has been used to measure λ_a of fiber glass batts and obtain data that show the effects of density, temperature, temperature difference, thickness, plate emittance, and heat flow direction on λ_a. A determinate error analysis indicates a λ_a measurement uncertainty of $\pm 1.7\%$. The system repeatability and reproducibility were found to be $\pm 0.2\%$. The measured performance of the tester agrees with that predicted by a thermal model. Analysis of the transient characteristics of the screen when it is subjected to a step change in power suggests that the apparatus can be used to determine apparent thermal diffusivity.

KEY WORDS: thermal conductivity, flat insulation tester, thermal modeling, total hemispherical emittance, screen heater, fiber glass insulation, thermal insulation, thermal resistance, unguarded technique

Nomenclature

- A Metering area, 0.610 by 0.914 m, 0.5575 m^2
- I Current through heater, amperes
- q Power, W

[1] Group leader, senior technologist, adjunct research, participant, and research staff member, Martin Marietta Energy Systems, Inc., Oak Ridge, TN 37830.

R Thermal resistance, m^2 K/W
T_1 Specimen cold face temperature, K or °C
T_2 Specimen hot face temperature, K or °C
T_m Mean temperature, $(T_2 + T_1)/2$, K or °C
V Voltage drop for 0.61 m, volts
ΔT Temperature difference for heat flow correction (Eq 3), K
ΔX Specimen thickness, m
λ Thermal conductivity, W/m K
λ_a Apparent thermal conductivity, W/m K
λ_B Apparent thermal conductivity of batt for heat flow correction (Eq 3), W/m K
ρ Density, kg/m^3

This paper describes the development of a simple steady-state technique to obtain one dimensional heat flow for measurement of the apparent thermal conductivity (λ_a) of flat specimens of thermal insulation. The λ_a of a homogeneous material at steady state with one-dimensional heat flow is defined by Fourier's law as

$$\frac{q}{A} = -\lambda \frac{dT}{dX} \simeq \lambda_a \frac{(T_2 - T_1)}{\Delta X} \tag{1}$$

The right hand side of Eq 1 is a useful approximation that is exact if λ_a is constant or linearly dependent on temperature. The λ_a of a material is a measure of its ability to conduct heat and hence provides a quantitative means to evaluate the thermal performance of insulating materials. The λ_a of insulating materials often depends on a number of variables including material type, density, temperature, heat-flow direction, bounding-surface emittances, thickness, and structure, none of which are explicitly given in Eq 1. A variety of steady-state and nonsteady-state techniques have been developed to measure the heat flow characteristics of specific thermal insulation materials. The low value of λ_a of most commercially available thermal insulations and the simultaneous transport of heat by several mechanisms make determinations of λ_a difficult, time consuming, expensive, often inaccurate, and sometimes dependent on the measurement technique.

The ASTM Committee C-16, Thermal Insulations [*1*], has fostered and encouraged the development of testing equipment and procedures to measure the λ_a of thermal insulations. These include techniques intended to obtain one-dimensional heat flow: the guarded hot plate [*2*], the heat flow meter [*3*], and the guarded and calibrated hot boxes [*4,5*] and techniques to obtain radial heat flow such as the pipe insulation tester [*6*]. Each of these is mechanically complex, expensive to construct and operate, and require great effort to obtain accurate data. Subtle difficulties of these methods are described by Pratt [*7*] and Bode [*8*].

The apparatus to be discussed uses a horizontal, unguarded, electrically heated flat nichrome screen wire heater as a heat source. The nichrome screen is sandwiched between two layers of insulation (two specimens) with flat, isothermal bounding surfaces. The low thermal conductivity of the nichrome screen reduces unwanted horizontal heat flow and thus provides the desired vertical, one-dimensional heat flow to the temperature-controlled, water-cooled copper plates that are the bounding surfaces. Other investigators [*9–13*] have noted that use of a large area and a flat, thin heater of poor thermal conductivity avoids the need for special guard-heater control systems in a hot plate apparatus and can provide uniform heat generation from a source with low heat capacity. The present design evolved from a pipe-insulation tester with an unguarded, cylindrical screen-wire heater [*14*], and from analysis and construction of prototype flat insulation testers with flat screen wire heat sources [*15,16*].

The magnitude of errors for this screen heater apparatus was assessed using a finite-difference heat-conduction program [*17*] known as HEATING5. HEATING5 can be used to solve heat conduction problems in Cartesian coordinates and can provide steady-state temperature distributions for several configurations of the apparatus in its normal mode of operation. This allows determination of the errors in the measured apparent thermal conductivity associated with the assumption of one-dimensional heat flow in a central region of the tester at steady state. Figure 1 is a schematic drawing of the three-dimensional system that was modeled to assess the system accuracy when operated without edge guarding. Specific geometries and properties used in the modeling are given in Table 1. The numerical calculations were used to study the effect of specimen and edge insulation thicknesses on the steady-state temperature distribution within the specimen. Figure 2 summarizes the resulting error as a function of distance from the screen center for two thicknesses of edge insulation and for three specimen thicknesses. These results show edge losses cause the apparent thermal conductivity to be always greater than the input thermal conductivity; that is, a positive error for "cold guards."

The results shown in Fig. 2 were obtained with constant thermal conductivities, specimen thicknesses of 0.0762, 0.1524, and 0.3048 m, and edge insulation thicknesses of 0.0, 0.0762, and 0.1524 m. For a 0.0762-m-thick specimen, the numerical results show a central region about 0.40 m square in which the edge loss error is less than 1% with no edge insulation, and edge loss error with 0.0762 or 0.1524 m of edge insulation is significantly below 1%. Results for 0.1524-m-thick specimens show the minimum edge loss error is over 1% even with 0.1524-m edge insulation. Calculations done for 0.3048-m-thick specimens show edge loss errors that are unacceptably large. The calculations with specimens up to 0.0762 m thick with $\lambda_a = 0.046$ W/m K show that measured thermal conductivity error due to edge losses are less than 0.5% if screen temperatures are measured within the 0.4 m central

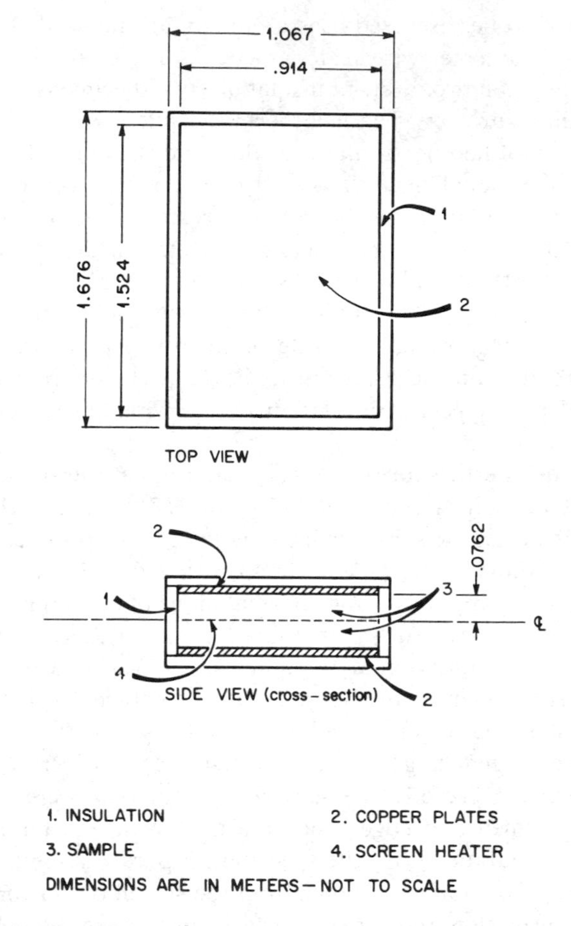

FIG. 1—*Three-dimensional schematic of system to be modeled.*

square and 0.0762 m of edge insulation is used. Clearly these calculations show the usefulness of pursuing this simple technique to obtain accurate values for λ_a.

Equipment Description

The assembled apparatus without perimeter insulation is shown in Fig. 3. This photograph shows the 1.1 by 1.7-m table with four threaded corner posts for holding the top and bottom copper plates and the screen frame in their respective horizontal positions. Proceeding upward from the table top one can see the lower cold plate insulation consisting of two insulating boards and a compressed fiber glass batt, the lower cold plate, the lower specimen, the screen heater frame, the upper specimen, the upper cold plate, and the upper

TABLE 1—*Finite difference thermal model dimensions and properties for one octant with 30°C faces.*

A. DIMENSIONS AND PROPERTIES

Item	Length, m	Width, m	Thickness, m	Thermal Conductivity, W/m K
Screen	0.457	0.762	3.175×10^{-4}	1.6 (Ref *15*)
Sample	0.457	0.762	See B	0.046

B. COMBINATIONS CALCULATED

Specimen Thickness, m	Edge Insulation Thickness, m		
	0.0	0.0762	0.1524
0.0762	X	X	X
0.1524		X	X
0.3048		X	X

C. VOLUMETRIC HEAT GENERATION RATE IN SCREEN: 0.0156 W/cm^3

D. PROGRAM CONVERGENCE LIMIT: 1×10^{-8}

cold plate insulation consisting of a compressed fiber glass batt and two insulating boards. The upper and lower cold plate insulation (0.11 m thick) provides an R-value of about 4.7 m^2 K/W (27 h · ft^2 °F/Btu). The two water circulating units for independent plate temperature control are shown below the table. This independent temperature control allows measurements to be made with heat flow up, heat flow down, or heat flow from the screen to both plates at the same temperature.

Figure 4 is a schematic drawing of the temperature-control plumbing for a plate. The solid bounding surfaces were fabricated from copper plate, 1.2 cm thick, 1.14 m wide, and 1.75 m long. Four holes are provided for the positioning posts and 1-cm-diameter copper tubing was soldered to the external surface of the copper plate for the circulating fluid. The copper tubing provided two counter current flow paths for water from the Braun water bath [*18*], with flow boosted by an external pump and balanced with two rotameters. The plate temperature profile and input/output temperatures were obtained using 15 Type E (Chromel P and constantan) 0.12-mm-diameter thermocouples that were held to the plate surface with thermally conducting epoxy [*19*]. The surfaces facing the specimen were machined to a flatness of ±0.13 mm, and six table support screws were adjusted to provide a flatness to this level. Initial tests were run with the as-machined surface facing the specimen and later both plates were painted with a black paint [*20*] to determine the effect of bounding surface emittance on the results.

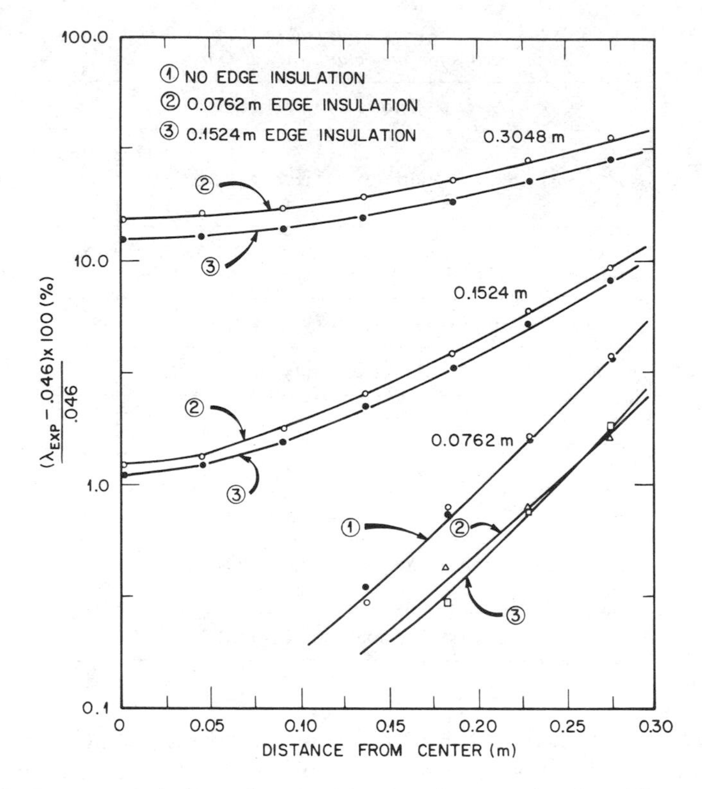

FIG. 2—*Percent error in thermal conductivity of specimen as a function of distance from the screen center for three specimen thicknesses and two levels of edge insulation.*

Figure 5 is a schematic drawing of the instrumented nichrome screen wire heater in its support frame. The screen [*21*] was made from 0.25-mm-diameter wire woven on a 40 by 40 mesh (per inch) and has a thickness of 0.64 mm, width of 0.91 m, and length of 1.52 m. This screen was held in the frame with 14 electrically insulated tension springs with adjusting nuts. Distributed current input leads joined copper bars brazed to each end of the heater. Four voltage taps were welded to the edge of the heater centered on the screen with separations of 0.610 and 0.914 m. Ten Type E 0.12-mm-diameter thermocouples were attached to the screen with thermally conducting epoxy on the 0 and 180° axis with 0.3 m spacing. The positions of the spacers used to maintain the specimen thickness are shown in Fig. 5.

The 40 Type E thermocouples of the apparatus join copper wires in a pair of ice-water reference baths. All thermal electromotive forces (emfs) were measured with an L&N Type K-5 potentiometer (1.6 V accuracy is $\pm(0.001\% \times V + 2\ \mu V)$) or an Autodata Ten Scanner (uncertainty $\pm 1\ \mu V$ in 30 000 μV). The electrical power to the screen heater was provided by a Hewlett Packard d.c. supply 6260 B (10 V, 100 A) with a standard 0.01 ohm

FIG. 3—*Photograph of assembled apparatus without edge insulation showing component positions.*

resistor used for current measurements. The voltage taps with a spacing of 0.610 m on the screen edge define the metered area of 0.610 by 0.914 m.

The specimen thickness is fixed by eight 3-cm-diameter micarta tube spacers positioned outside the 0.610 m voltage tap spacing. Four such spacers are positioned on each side of the screen and spacer tube lengths from 4 to 9 cm ± 0.001 cm are available.

Equipment Characteristics and Test Procedure

Several auxiliary tests were conducted in the course of establishing the operational characteristics and test procedures for the screen heater apparatus.

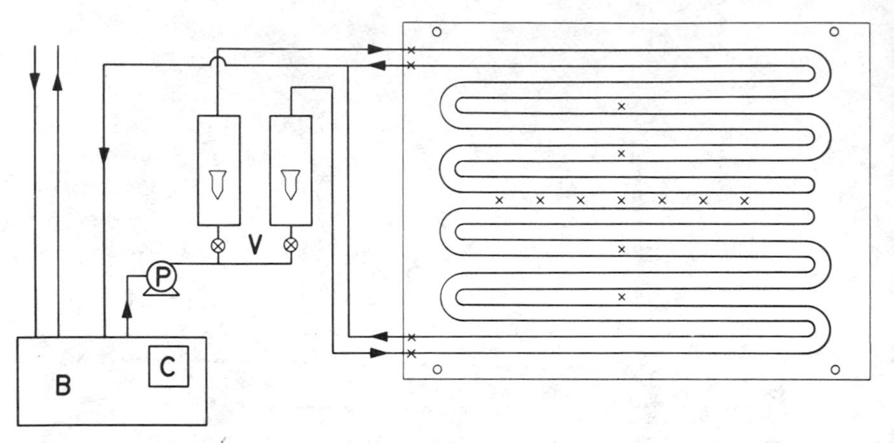

FIG. 4—*Schematic drawing of the temperature control and plumbing systems for each cold plate showing dual counter current flow path, Braun water bath (B) and controller (C), booster pump (P), balancing valves (V), and 15 thermocouple positions (x).*

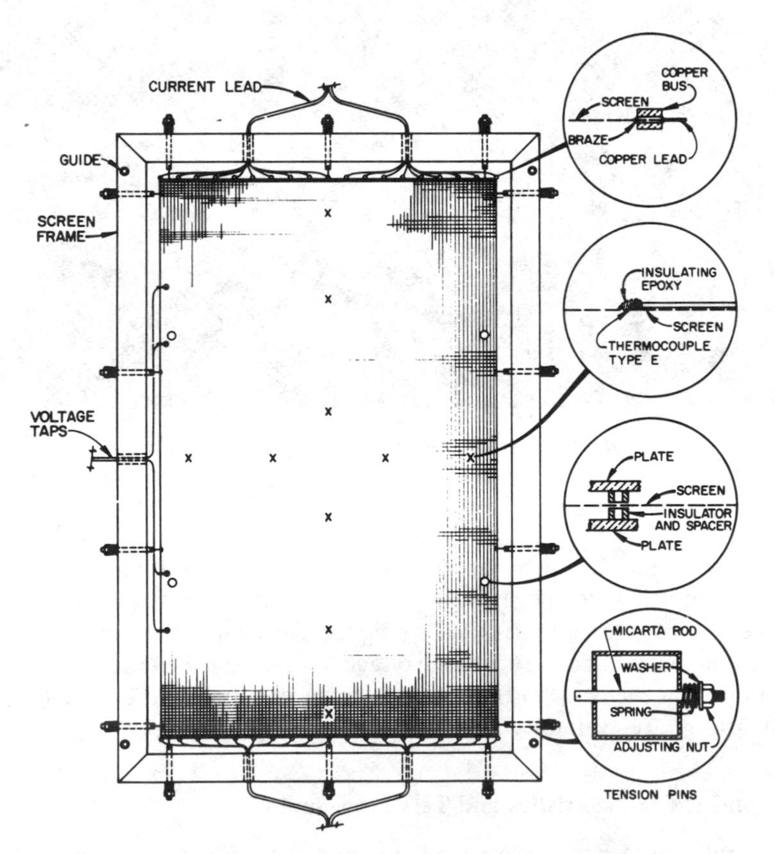

FIG. 5—*Schematic drawing of instrumented nichrome screen wire heater in support frame with distributed current leads, voltage taps, specimen thickness spacers, tension pins, and thermocouple positions (x).*

Figure 6 shows that the cold plate temperature uniformity near 30°C was better than ±0.03°C when the plate had 7.6 cm of insulation on both sides with an R-value of 3.7 m^2 K/W (21 h · ft^2 °F/Btu). The 11 Type E thermocouples are distributed over an area, 0.6 by 0.9 m and agreed to ±2 μV. A portion of this may be due to edge and circulation losses.

Two coats of 3M-SCS-2200 paint [*20*] were used to blacken the cold plates during the initial tests. This type paint replaces the previously produced and recommended coating for plates in guarded hot plate (GHP) [*2*] and heat flow meter (HFM) [*3*] apparatus. The total hemispherical emittance [*22*] of this paint was determined in vacuum to 150°C using an instrumented electrically heated strip. The results are shown in Fig. 7 and indicate an emittance value of 0.75 at 50°C, which is significantly lower than the recommended value of 0.95 for the plates [*2*].

A nominal R-11 fiber glass batt was compressed to a thickness of 7.62 cm using the micarta spacers to fix the plate-screen-plate spacing. Perimeter insulation of about 7 cm thickness was installed, and the cold plates were brought to the same average temperature near 30°C. After equilibration and prior to applying electrical power to the screen, the 10 thermocouples on the screen were read with the results shown in Fig. 8. This shows temperature uniformity of ±0.02°C for an area, 0.3 by 0.9 m and ±0.17°C for a larger area of 0.7 by 1.4 m. Indeed the resulting temperature profile agrees with the results from the thermal model and shows the effects of heat losses along the unguarded edges.

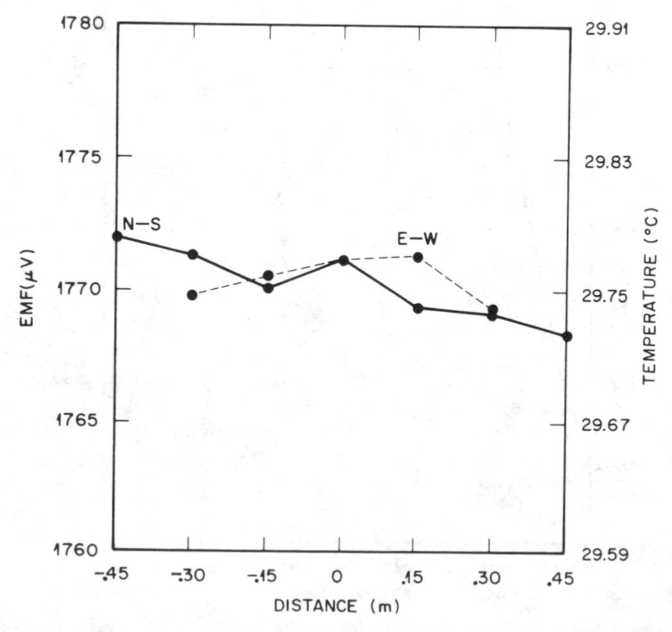

FIG. 6—*Cold plate temperature distribution near 30°C with both sides insulated. Type E thermocouple sensitivity is 61.03 μV per degree at 30°C. N-S is longer dimension of apparatus.*

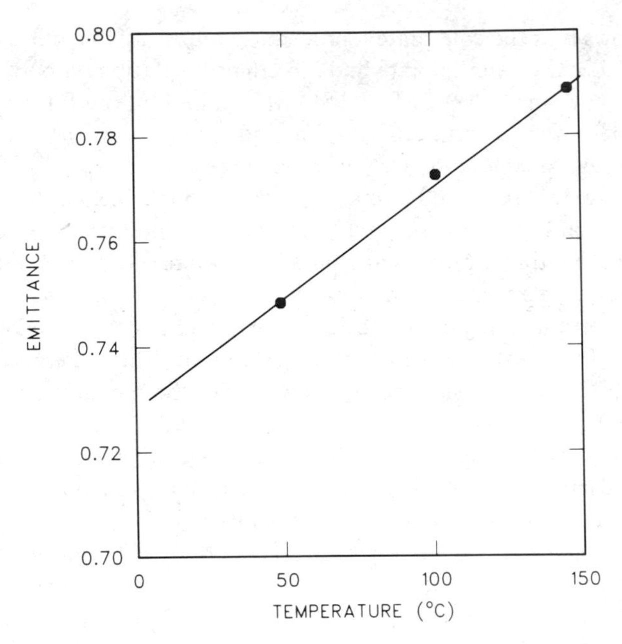

FIG. 7—*The total hemispherical emittance of two coats of 3M-SCS-2200 as a function of temperature.*

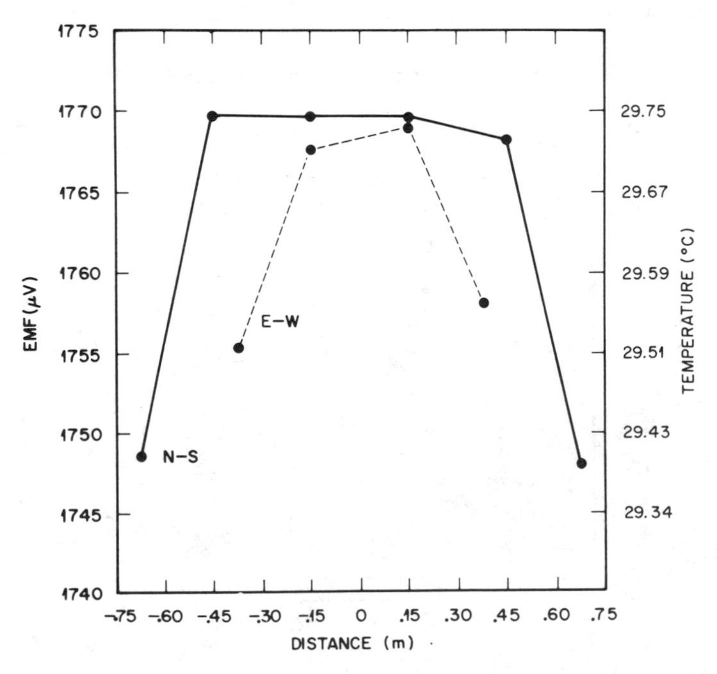

FIG. 8—*Screen temperature distribution with no power applied and plates near 30°C. N-S is longer dimension of apparatus.*

The first test with power applied to the screen had the cold plates near 30°C, a screen center temperature near 40°C, two-sided heat flow, and used 7.62-cm-thick fiber glass batts. Figure 9 shows the thermal emfs observed at steady state as a function of position on the screen, designated 0° orientation for the batt specimen. As expected, the screen temperatures are highest near the central portion of the screen and lower toward the screen edges due to edge heat losses. Unexpectedly the central four thermocouples exhibited a much greater spread than indicated by the zero-power test results shown in Fig. 8. The cause for this was suspected to be due to local density variations in the specimen batts near the four central thermocouples which generated different local thermal resistances between each sensor and the cold plates. To test this hypothesis, the two central specimen batts were rotated 180° in the plane of the screen. The central four thermocouples showed a corresponding change in value with orientation as shown in Fig. 9.

The two batts on either side of the central portion of the screen were removed, and four 15-cm-diameter disks were cut from each at positions corresponding to the screen thermocouples. Table 2 shows the weights of pairs of disks having a range of about 20%. The corresponding emf range of about 40 μV (2/3°C) for a specimen ΔT of 10° suggests about a 7% change in λ_a. This is very consistent with the known change in λ_a with density and provides an

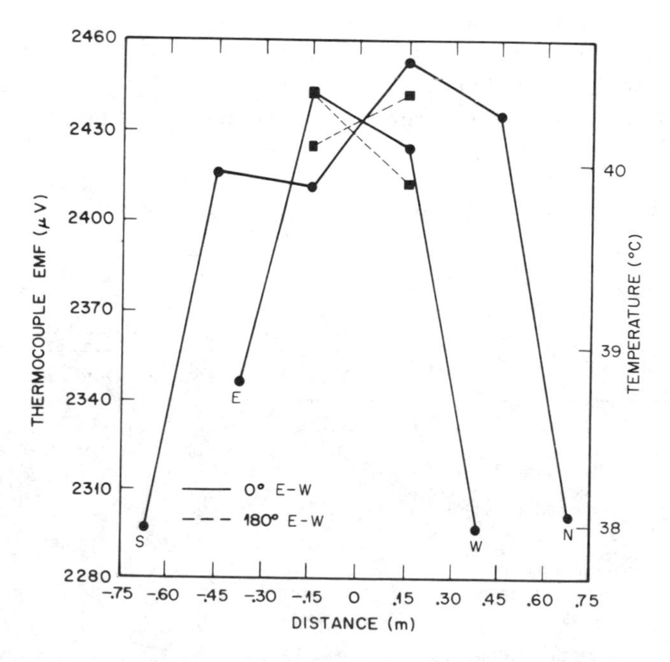

FIG. 9—*Screen temperature distribution with power applied, plates near 30°C, and for two specimen orientations.*

TABLE 2—*Weights of pairs of 15-cm-diameter disks cut at thermocouple positions.*

Thermocouple Position	Disk Weight, g
East	30.0
North	29.9
South	24.5
West	27.8

added advantage for the screen heater system. This suggests the current screen thermometry is adequate for obtaining average thermal performance.

Table 3 provides λ_a-values obtained for this specimen for four conditions of heat flow for a mean temperature of 35°C with a temperature difference of 10.7°C. The two-sided values agree to 0.13% when the specimen was rotated 180°, and the average of the two-sided and one-sided heat flow tests agree to 0.17%.

The foregoing tests yielded a useful procedure to obtain the measurables required by Eq 1 for the apparent thermal conductivity of low-density fiber glass specimens. These results suggested that the apparatus performed as expected. In these calculations the average screen temperature was based on the four central thermocouples positioned on a diamond pattern with 0.3 m spacing. The agreement of the 30 plate thermocouples indicated the central thermocouple on each plate could provide the cold boundary temperatures. Such potentiometric data were recorded as a function of time to validate that system temperatures were not changing with time. The following equations were used to calculate λ_a from experimentally measured quantities.

TABLE 3—*Apparent thermal conductivity values for four heat flow states for a fiber glass batt.*

Heat Flow State	Apparent Thermal Conductivity, $\times 10^2$ (W/m K)	Temperature, °C	
		Mean	Difference
Two-sided	4.728	35.15	10.74
Two-sided (180° rotation)	4.734	35.12	10.72
Two-sided average	4.731		
One-sided, down (bottom batt)	4.763	35.26	10.96
One-sided, up (top batt)	4.716	35.26	10.72
One-sided average	4.739		

Two-sided heat flow

$$\lambda_a = \frac{I \cdot V[\Delta X(\text{micarta})]}{2A[T_{(\text{screen})} - T_{(\text{plates})}]} \tag{2}$$

One-sided heat flow (with heat flow corrections)

$$\lambda_a = \frac{\left[I \cdot V - \left(\lambda_B \cdot A \cdot \frac{\Delta T(B)}{\Delta X}\right)\right][\Delta X]}{A[T_{(\text{screen})} - T_{(\text{plate})}]} \tag{3}$$

Results

Specimen A was a fiber glass batt, specimen code 1101-1, that had been previously tested in a GHP apparatus [*23*]. This batt was 0.62 m wide, and two test strips 1.15 m long were weighed to obtain an average weight per area of 0.861 kg/m^2. The apparent thermal conductivity of Specimen A was measured at three thicknesses obtained by compression, at four mean temperatures, and with the bounding plate surfaces unpainted and painted with two coats of 3M-SCS-2200. Figure 10 shows the λ_a-values are increased about 2% by increasing the cold plate emittance from that of a machined copper surface to 0.75. The λ_a for the 0.75 emittance surfaces increase with mean tempera-

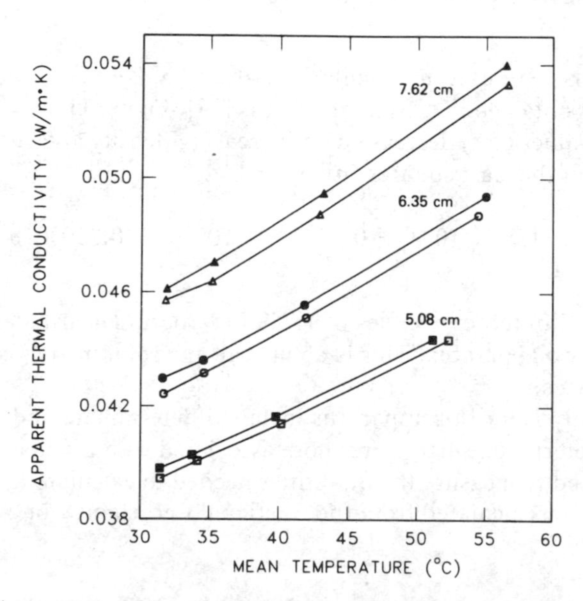

FIG. 10—*Apparent thermal conductivity as a function of mean temperature for a fiber glass batt at three thicknesses. The open symbols are for bare machined copper cold plate surfaces and the filled symbols are for painted copper cold plate surfaces.*

ture and decrease with increasing density, which occurred as the specimen was compressed. A variety of data representations have been suggested for such results. For example, it has been shown that for coupled conductive and radiative heat transport in semitransparent materials, with absorption coefficients near 300 m^{-1}, λ_a can be represented by linear relation with mean temperature to better than 1% [*24*]. The use of a linear mean temperature representation may be criticized on the basis that contributing radiative heat transport should vary as the absolute temperature cubed. An attempt to include a physically real representation of the λ_a results was made by first subtracting a linear equation for the thermal conductivity of air [*25*] and fitting the remainder with a constant and a cubic temperature term. The results for three density levels are given by Eqs 4 to 6, which had a maximum standard deviation of 0.6×10^{-4}

for 11.28 kg/m^3,

$$\lambda_a = 0.13093 \times 10^{-2} + 0.69073 \times 10^{-4} T_m + 0.83407 \times 10^{-9} T_m^3 \quad (4)$$

for 13.54 kg/m^3,

$$\lambda_a = 0.23126 \times 10^{-2} + 0.69073 \times 10^{-4} T_m + 0.68968 \times 10^{-9} T_m^3 \quad (5)$$

for 16.92 kg/m^3,

$$\lambda_a = 0.31345 \times 10^{-2} + 0.69073 \times 10^{-4} T_m + 0.55197 \times 10^{-9} T_m^3 \quad (6)$$

These equations were used to calculate λ_a at 297 K (24°C) for comparison with five values obtained in a GHP apparatus [*23*]. Figure 11 shows the eight 297 K (24°C) values of λ_a decrease with increasing density and the deviation of the data from the least squares fit

$$\lambda_a = 0.23185 \times 10^{-1} + 0.75743 \times 10^{-4} \rho + 0.23039/\rho \quad (7)$$

The maximum difference in values of 1.7% in λ_a occurs near a density of 12 kg/m^3 for the two apparatus. This is about half the combined uncertainty of the two apparatus.

An error analysis for this apparatus included determinate and indeterminate errors. Determinate errors are those associated with the inaccuracies of instruments used to measure the quantities needed to calculate λ_a. Determinate errors can be calculated from the fractional uncertainty for λ_a, $\Delta\lambda_a/\lambda_a$, as follows

$$\left|\frac{\Delta\lambda_a}{\lambda_a}\right| = \pm\left|\frac{\Delta q}{q}\right| \pm \left|\frac{\Delta A}{A}\right| \pm \left|\frac{\Delta(\Delta X)}{\Delta X}\right| \pm \left|\frac{\Delta(\Delta T)}{\Delta T}\right| \pm \left|\frac{d\lambda}{dT}\,\frac{\Delta T}{\lambda}\right| \quad (8)$$

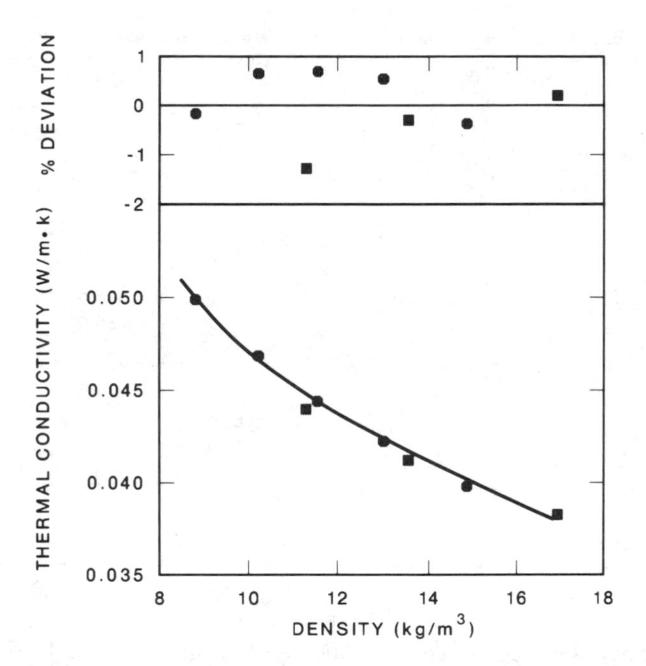

FIG. 11—*Apparent thermal conductivity at 24°C as a function of density for a fiber glass batt as measured by a guarded hot plate* (●) *and the unguarded screen apparatus* (■). *Percent deviation from* $\lambda_a = a + b\rho + c/\rho$.

Indeterminate errors are based on deviation of the system from assumed boundary conditions and may be estimated.

Table 4 provides the basis for calculating the fractional uncertainty for the quantities given in Eq 8, and the total uncertainty is ±1.7% for an assumed temperature difference of 10°C. The individual components in this total uncertainty include:

(*a*) dimensional uncertainty of ±0.39%: a micrometre error for 7.6-cm-long micarta spacers of ±0.007 and ±0.013 cm for plate flatness yields ±0.26%; ruler error of ±0.05 cm for the voltage tap separation of 61 cm and the screen width of 91 cm that yields ±0.13%;

(*b*) power uncertainty of ±0.053%: the K-5 potentiometer accuracy on the 1.6 V range and stated equipment accuracies for the voltage divider, the standard resistor, and the d.c. power supply; and,

(*c*) the temperature error of ±0.058°C for a 10°C temperature difference or ±1.16%: a platinum resistance thermometer calibration error, the potentiometer measurement error and observed wire-to-wire differences on the plate and screen with no heat flow.

TABLE 4—*Fractional uncertainties for the unguarded screen tester.*

	Value	Error	Error, %	
Thickness, cm	7.6	0.020		0.26
Length, cm	61	0.05	0.08	
Width, cm	91	0.05	0.05	
Area				0.13
Voltage, V	0.4	6×10^{-6}	0.0015	
Current, A	0.4	6×10^{-6}	0.0015	
Voltage divider			0.02	
Standard resistor			0.01	
Power supply			0.02	
Power				0.053
Temperature, K				
Calibration		0.039		
Potentiometer		0.003		
Wire to wire		0.016		
Difference	10	0.058		1.16
Property/degree	0.006	0.12		0.072
Total uncertainty				1.675
Root of sum of squares				1.199

The change of λ_a per degree is about 0.006 K^{-1}, so an error of 0.12 K contributes ±0.072% uncertainty. The sum of the fractional uncertainties calculated from Eq 8 is ±1.7%, while the most probable uncertainty (the square root of the sum of the squares) is about ±1.2%.

The temperature difference selected in this analysis and others [*26*] is a dominant term. The 10°C difference chosen for the present analysis is less than the value normally associated with ASTM methods.

The indeterminate errors include items such as system variation from assumed boundary conditions and those from sources that may be difficult to eliminate. The thermal modeling of edge heat losses shows the λ_a (measured) is about 0.2% greater than λ_a (true) for a 7.6-cm-thick fiber glass batt if adequate edge insulation is provided. Inhomogenieties in a specimen have been demonstrated to change λ_a. Nonuniformity in the screen wire heater such as weave, or width could change λ_a. Increasing the emittance of the bounding surfaces to 0.75 has been shown to increase λ_a, and, if they could be made to have an emittance of one, further increases would be expected. One way to estimate the indeterminate errors is to test the system with a reference material and this is planned.

Thermal Diffusivity Determination

Transient temperature distributions resulting from a step change in screen power can be calculated with HEATING5 [*17*] for specified thermal conductivities, specific heats, and densities for the screen and test material. The calculated temperature distribution includes values for the screen temperature as a function of time which can be compared with experimental screen temperatures. The specimen properties are adjusted to bring calculated and experimental screen temperatures into agreement.

Figure 12 shows a comparison of the measured screen temperatures and curves calculated with thermal diffusivities of 0.0256 and 0.0210 cm^2/s. These are about one third of the diffusivity calculated using the steady-state λ_a, and published heat capacity and true density data [*27*]. These data were obtained with a 0.0635-m-thick specimen of fiber glass batt insulation. Electrical power to the tester was sufficient to raise the screen temperature from approximately 29.7 to 39.8°C. This analysis was done to demonstrate the feasibility of using the screen tester to determine a transient property. This significant deviation of thermal diffusivity and its cause will be the subject of later papers.

Conclusions

The feasibility of accurately measuring λ_a with an unguarded longitudinal heat flow technique that uses an electrically heated nichrome screen has been demonstrated.

This simple unguarded technique has a most probable error of ±1.2% in the range 300 to 330 K and results with heat flow up, heat flow down, or heat flow to both plates agreed to ±0.2%. Agreement of λ_a for a fiber glass batt with previous measurements by a GHP was better than ±2%. Specific effects on λ_a of density, temperature, temperature difference, and plate emittance were determined for this specimen. This system may be useful to determine apparent thermal diffusivity, but yields significantly lower values than expected.

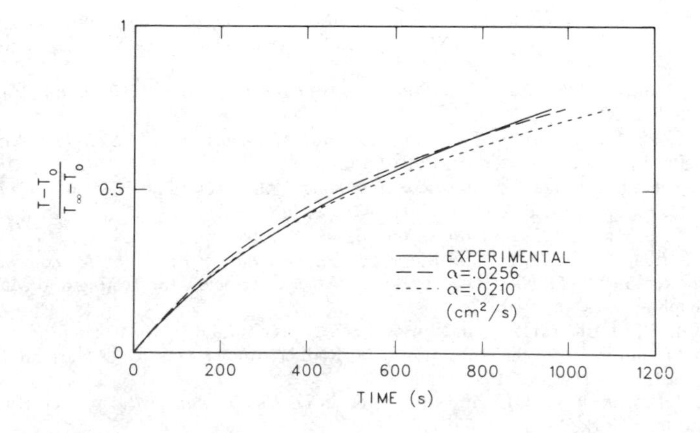

FIG. 12—*Comparison of measured and calculated screen heater temperatures for a step change in screen power.*

Acknowledgments

Appreciation is extended to D. Kiplinger for his key role in helping design and construct the screen heater apparatus. The paper was reviewed by T. G. Godfrey, G. L. Copeland, and T. S. Lundy, whose comments were most helpful and improved our description of this system. We are most appreciative of the manuscript typing and preparation by B. Hickey, C. Whitus, and T. L. O'Kelly. This project was a subtask of the U.S. Department of Energy Program for Building Thermal Envelope Systems and Insulating Materials.

The research was sponsored by the Office of Building Energy Research and Development, U.S. Department of Energy, under contract DE-AC05-840R21400 with the Martin Marietta Energy Systems, Inc.

References

[1] American Society for Testing and Materials Committee C-16, Thermal Insulations, C-16 Chairman, J. D. McAllister, 1916 Race Street, Philadelphia, PA, 19103.
[2] "Standard Test Method for Steady-State Thermal Transmission Properties by Means of the Guarded Hot Plate," ASTM C 177-76, 1982 Annual Book of Standards, American Society for Testing and Materials, Philadelphia, Part 18, pp. 20-53.
[3] "Standard Test Method for Steady-State Thermal Transmission Properties by Means of the Heat Flow Meter," ASTM C 518-76, 1982 Annual Book of Standards, American Society for Testing and Materials, Philadelphia, Part 18, pp. 222-253.
[4] "Standard Test Method for Steady-State Thermal Performance of Building Assemblies by Means of a Guarded Hot Box," ASTM C 236-80, 1982 Annual Book of Standards, American Society for Testing and Materials, Philadelphia, Part 18, pp. 81-92.
[5] "Test Method for the Calibrated Hot Box," ASTM C 976-83, to be published in the 1983 Annual Book of Standards, American Society for Testing and Materials, Philadelphia, Part 18.
[6] "Standard Test Method for Steady-State Heat Transfer Properties of Horizontal Pipe Insulations," ASTM C 335-79, 1982 Annual Book of Standards, American Society for Testing and Materials, Part 18, pp. 100-113.
[7] Pratt, A. W. in *Thermal Conductivity, Volume 1*, R. P. Tye, Ed., Academic Press, New York, 1969, Chapter 6, pp. 301-405.
[8] Bode, K.-H., *International Journal of Heat Mass Transfer*, Vol. 23, 1980, pp. 961-970, also available as translated in ORNL/tr-4932 (May 1983) Oak Ridge National Laboratory, Oak Ridge, TN.
[9] Niven, C. and Geddes, A. E. M., *Proceedings of the Royal Society*, London, Vol. A87, 1912, pp. 535-539.
[10] Gilbo, C. F., *Symposium on Thermal Insulating Materials, ASTM STP 119*, American Society for Testing and Materials, Philadelphia, 1951, p. 45.
[11] Hager, N. E., Jr., *Review of Scientific Instruments*, Vol. 31, No. 2, Feb. 1960, pp. 177-185.
[12] Hager, N. E., Jr., U.S. Patent No. 3,045,473, 24 July 1962.
[13] Hager, N. E., Jr., *ISA Transactions*, Vol. 8, No. 2, 1969, pp. 104-109.
[14] Jury, S. H., McElroy, D. L., and Moore, J. P., *Thermal Transmission Measurements of Insulation, ASTM STP 660*, R. P. Tye, Ed., American Society for Testing and Materials, Philadelphia, 1978, pp. 310-326.
[15] Moore, J. P., McElroy, D. L., and Jury, S. H., "A Technique for Measuring the Apparent Thermal Conductivity of Flat Insulations," ORNL/TM-6494, Oak Ridge National Laboratory, Oak Ridge, TN, Oct. 1979.
[16] Moore, J. P., McElroy, D. L., and Jury, S. H. in *Thermal Conductivity 17*, J. G. Hust, Ed., Plenum, New York, 1983, pp. 727-735.
[17] Turner, W. D., Elrod, D. C., and Siam-Tov, I. I., "HEATING5—An IBM 360 Heat Conduction Program," ORNL/CSD/TM-15, Oak Ridge National Laboratory, Oak Ridge, TN, March 1977.

[*18*] B. Braun Instruments, Thermomix 1480 and Frigomix 1495, San Mateo, CA.
[*19*] Astrodyne, Inc., Thermal Bond 312, Burlington, MA.
[*20*] Energy Control Products Projects 3M-SCS-2200, Experimental Solar Absorber Coating, St. Paul, MN 55144.
[*21*] Phoenix Wire Cloth, Inc. 40 × 40 Per Inch Nichrome V Wire, 0.010 inch, P.O. Box 610, Troy, MI 48084.
[*22*] McElroy, D. L. and Kollie, T. G. in *Measurements of Thermal Radiation Properties of Solids, NASA SP31*, National Aeronautics and Space Administration, J. C. Richmond, Ed., 1963, pp. 365-379.
[*23*] Tye, R. P., Desjarlais, A. O., Yarbrough, D. W., and McElroy, D. L., "An Experimental Study of Thermal Resistance Values (*R*-Values) of Low-Density Mineral-Fiber Building Insulation Batts Commercially Available in 1977," ORNL/TM-7266, Oak Ridge National Laboratory, Oak Ridge, TN, April 1980.
[*24*] Fine, H. A., Jury, S. H., McElroy, D. L., and Yarbrough, D. W. in *Thermal Conductivity 17*, J. G. Hust, Ed., Plenum, New York, 1983, pp. 359-368.
[*25*] Touloukian, Y. S., Liley, P. E., and Saxena, S. C., *Thermophysical Properties of Matter, Volume* 3, Plenum, New York, 1970, pp. 512.
[*26*] Siu, M. C. I. and Bulik, C., *Review of Scientific Instruments*, Vol. 52, No. 11, 1981, pp. 1709-1716.
[*27*] Wilkes, K. E., *Thermal Performance of Exterior Envelopes of Buildings II*, ASHRAE SP 38, American Society of Heating, Refrigerating, and Air-Conditioning Engineers, Inc., 1983, pp. 131-159.

Mark Bomberg[1] *and K. Richard Solvason*[1]

Discussion of Heat Flow Meter Apparatus and Transfer Standards Used for Error Analysis

REFERENCE: Bomberg, M. and Solvason, K. R., **"Discussion of Heat Flow Meter Apparatus and Transfer Standards Used for Error Analysis,"** *Guarded Hot Plate and Heat Flow Meter Methodology, ASTM STP 879*, C. J. Shirtliffe and R. P. Tye, Eds., American Society for Testing and Materials, Philadelphia, 1985, pp. 140–153.

ABSTRACT: The need for thermal resistance measurements on thick specimens (100 to 180 mm) of low-density thermal insulation and subsequent changes in ASTM test standards have increased the need for the evaluation of precision and accuracy attainable with the heat flow meter (HFM) apparatus. Such an evaluation is usually performed with transfer standards established on a guarded hot plate. The Division of Building Research of the National Research Council of Canada (DBR/NRCC) has developed a transfer standard and calibrated specimen bank to be used in verifying HMF test procedures and instrument error evaluation. This paper discusses such calibration and the use of the HFM apparatus and reviews the transfer standards developed at NRCC.

KEY WORDS: thermal resistance, thermal conductivity, heat flow measurements, heat flow transducers, heat flow apparatus, heat flow meter

For the past several years the only Canadian laboratory performing thermal conductivity (thermal resistance) tests on insulating materials for official evaluation and acceptance programs has been that of the Division of Building Research of the National Research Council of Canada (DBR/NRCC). It has been also responsible for developing expertise in thermal conductivity testing.

Rapid developments in thermal insulating materials and advancement in laboratory instrumentation has forced DBR/NRCC to review its role. Technology transfer, enhancement of the capability of industrial laboratories to perform reliable and precise measurements, has been deemed more important than the maintenance of one, highly specialized, national testing facility.

[1]Senior research officers, Division of Building Research, National Research Council of Canada, Ottawa, Canada.

This paper reviews recent developments and, in particular, information developed at DBR/NRCC related to:

1. Determination of uncertainty in thermal resistance testing, using transfer standard specimens developed at DBR/NRCC.
2. Determination of the technical expertise or proficiency of laboratory staff, using calibrated specimens developed at DBR.

The term "transfer standard" is used to describe an insulation specimen with an established mean *R*-value and an uncertainty estimate of its thermal resistance. The term "calibrated specimen" is used to describe a specimen that is sent to a laboratory for a proficiency check. The laboratory must determine the thermal resistance of the specimen, and the result is then compared, by the certifying agency, with the characteristic obtained by the standard laboratory, in this case, DBR/NRCC.

The two apparatuses in most common use for laboratory determination of thermal resistance are the guarded hot plate (GHP) and the heat flow meter (HFM). At DBR/NRCC the GHP apparatus is used to develop transfer standards and calibrated insulation specimens; the HFM apparatus is used for routine thermal conductivity testing. It is also used by most independent commercial testing laboratories and by the research and development and quality control laboratories of manufacturers of thermal insulating material.

In this discussion a HFM apparatus means an instrument for laboratory determination of thermal resistance that employs one or more heat flow transducers. Although similar comparative methods have existed for approximately 100 years, it is only recently that the range of testing conditions in which the HFM method is utilized has exceeded that of other laboratory techniques. Broadening the range of HFM apparatus applications, however, has introduced new problems.

Historic Background[2]

Between 1870 and 1878 Péclet proposed three different methods of determining thermal conductivity: sphere, pipe section, and a vertical slab. At about the same time, in 1872, Forbes introduced a slab method for determining negative temperatures in which he measured the thickness of the ice layer formed on one surface of the specimen while the other side was exposed to a freezing mixture. In 1881 Christiansen built a comparative instrument that later was developed into a HFM apparatus. It contained three relatively thick copper plates with carefully drilled holes for thermometers. Two specimens,

[2]Two Russian books, namely: B. S. Pietuchow, *Optnoje Izuczenije Procesow Tieplopieriedaczi (Experimental Study of Heat Transfer Processes)*, Gosenergoizdat, 1952, and D. N. Timrot, *Opriedielenije Tieploprowodnosti Stroitielnych Materialow (Determination of Thermal Conductivity of Building Materials)* Gosenergoizdat, 1932, provide a detailed review of the historic background for this area of testing.

one with known thermal conductivity and another for which thermal conductivity was to be measured, were placed between the plates. Christiansen proposed a formula

$$\lambda = \lambda_0 \frac{L\,(T_3 - T_2)}{L_0\,(T_2 - T_1)}$$

where

λ, L = thermal conductivity, thickness of the specimen,
λ_0, L_0 = same as above, for the reference material, and
T_3, T_2, T_1 = temperatures of copper plates, starting from the hot plate.

Over the years the slab apparatus has been developed into a GHP through the efforts of many research workers, for example, Lees in 1892, Lees and Chorlton in 1896, Niven in 1906, Nusselt and Groeber in 1911, and Pönsgen in 1912 who introduced guarding heaters surrounding the main heater.

The concept of a heat flow transducer was used in Germany by Hencky in 1919 and by Schmidt in 1923 for field work, particularly on industrial insulations. While Hencky added a layer with thermal resistance comparable to that being measured, Schmidt's transducer had a small thermal resistance and was provided with compensating rubber strips of approximately the same resistance as the transducer. Depending on the nature of the substrate and the edge loss error, the number of compensating (guarding) strips could be increased (for example, to five if the heat flow transducer was placed on the metal surface).

It may be surprising that 60 years later the same technique is used, with the same magnitude of error. The explanation is simple: the physical principles involved remain practically unchanged, but understanding of the measurement uncertainties has improved significantly. Progress in such measurements is therefore not so much in the development of more precise instruments as in enlarging the range of materials and testing conditions for a given level of precision and accuracy. HFM apparatus developments are very characteristic in this respect. With the abundance of inexpensive electronic components it is possible to build today an apparatus that can yield results several times faster than a GHP apparatus can, one moreover that permits testing under conditions in which the GHP would yield excessive errors.

Advantages and Drawbacks of the HFM Method

The HFM method is not an absolute one and requires calibration with specimens having both mean thermal resistance and the uncertainty range of this value determined by an absolute method, for example, by the GHP method. In an earlier paper [*1*] that compared various calibration techniques for the HFM the authors concluded that the preferred technique is to use transfer standards. As errors in measurement vary with thickness, mean temperature, and often even with temperature gradient across the specimen, one must either calibrate the HFM apparatus with the same type of specimen under the same testing conditions for which the apparatus will be used or perform a complex error verification to establish how different test conditions affect the uncertainty of HFM results.

If the HFM apparatus is to be used for quality control in the manufacture of thermal insulation, the calibration should be performed using a transfer standard of the same material as that manufactured. If it is to be used in a broad range of testing conditions, the same approach, that is, use of transfer standards, would require that GHP tests be performed on a variety of materials and under a variety of testing conditions. However, there may be significantly larger errors in GHP measurement results.

Another paper [*2*] discussing errors associated with GHP measurements dealt only with errors in the basic measurement characteristics of the equipment; the effect of the characteristics of the material under test was not discussed. Precision of the GHP apparatus was shown to be strongly dependent on the uniformity of heat flow across the gap between the metering and guard ring areas and "zero balance" of the system. It was also shown that only homogeneous, thin, uniform, "ideal" specimens could be used for experimental determination of the uncertainties in GHP testing.

Errors associated with the GHP apparatus arise from uncertainties in basic electrical (power and temperature) and thickness measurements and from deviation from one-dimensional heat flow [*3,4*]. The errors in basic measurements are characteristic of the equipment used; for example, for a 600 by 600-mm GHP apparatus at DBR/NRCC the repeatability precision of 0.2% and accuracy of 1.2% were determined at the 90% probability level. The errors resulting from multidimensional heat flow, that is, unbalance and edge loss error, depend on the nature of the material as well as on the equipment used and can be easily larger.

It is therefore important to remember that while the GHP method yields high precision and accuracy under well-defined conditions of unidirectional heat flow, this may not be the case for thick, low-density thermal insulations with a significant fraction of radiative heat transfer or for layered specimens with a high conductivity facing material.

Restrictions in the use of the GHP method relate to combinations of specimen thickness and testing conditions, where a lateral flow occurs over the air gap between main and guard ring heaters and in the specimen. Although a proper design of GHP apparatus may enlarge the range of testing conditions and material selection, its precision is reduced for thick specimens of thermal insulation faced with high-conductivity skins or membranes, layered and highly nonisotropic materials, or low-density materials in which a significant fraction of heat transfer is by radiation. In such cases the HFM apparatus may be used to better advantage than the GHP apparatus.

All these considerations limit the choice of specimen thickness and the materials and testing conditions that can be used for calibration and verification of HFM apparatus. If thicker transfer standards are needed, they must consist of two or more layers, each tested separately.

Principles of HFM Calibration

Bomberg and Solvason [*2*] have compared two methods of calibration. The first involves the use of calibrated specimens whose thermal resistance has been determined previously in the GHP. The heat flux in the HFM apparatus may be then inferred from the temperature difference across the specimen. This method is referred to as calibration of the HFM apparatus. In the second method, referred to as calibration of the heat flow transducer, the transducer and a dummy sample are placed between the plates of the one-sided GHP apparatus. The heat flux through the metering area of the transducer is assumed to be the same as that for the GHP measurement.

Both methods have advantages and disadvantages. In the calibration of the transducer, heat flow rate is measured directly and does not depend on a reference specimen. Accuracy depends, however, on the assumption that the heat flux through the dummy specimen and the transducer is strickly one-dimensional.

Two transducer types were examined: cork core [*5*], called ARM, and commercially available transducers [*6*] called SG heat flow meters. The 300-mm square transducers (ARM 1 and ARM 2) were built on a 6.3-mm core of Armstrong cork (No. 975), average density 496 kg/m^3, following the construction method described by Zabawsky [*7*]. Nine pairs of Chromel-Constantan thermocouple junctions were installed to provide a central 150 by 150-mm metering area. The total thickness of the transducer, including cork cover sheets, copper, and Mylar films, is about 8.4 mm. The surfaces were covered with the same black Nextel paint as was used on the SG heat flow meters, yielding the same emittance.

Table 1 gives the results of the calibration of ARM type transducers with high-density glass fiber and polystyrene transfer standards. Table 2 shows the

TABLE 1—*Calibration coefficients and confidence intervals as described by three standard deviations for ARM transducers (10 measurements).*

Calibration Procedure and Specimens	ARM 1, Hot Side		ARM 2, Cold Side	
	C_1 at 35°C, 3 σ		C_2 at 13°C, 3 σ	
Double HFM apparatus heat flux from high-density glass fiber and polystyrene transfer standards	25.03	0.30	24.69	0.48

TABLE 2—*Calibration coefficient of ARM 1 and ARM 2 heat flow meters as affected by change of layer adjacent to surfaces (heat flux determined from GHP apparatus).*

Test No.	Code	Materials in Contact with HFM	Heat Flux, W/m²	ARM 1		ARM 2	
				t_m, °C	C_1, W/m² mV	t_m, °C	C_2, W/m² mV
1	345-81	hot and cold plate of GHP	219.9	24.1	27.76	24.3	27.07
2	345-85	Hot plate, 12-mm RTV rubber	77.4	26.6	27.70	26.6	27.11
3	345-87	3-mm cork, 12-mm RTV rubber	56.6	24.6	26.20	24.7	25.72

calibration coefficient for the same transducers inserted in the square 300-mm GHP apparatus:

ARM transducer alone, Test 1,
ARM transducer in series with 12-mm RTV rubber, Test 2,
ARM transducer between 3-mm cork and 12-mm RTB rubber, Test 3.

The thermal properties of the layer adjacent to the transducer surface, as shown in Table 2, have a large effect on the apparent calibration coefficient, indicating a significant lateral component of heat flow.

Thus, one should calibrate the HFM apparatus using transfer standards, not the transducer alone. The transfer standards should be tested in the GHP apparatus under conditions identical to those used during HFM calibration and should cover the range of test conditions for which the HFM apparatus is intended. Most errors in calibration are, however, inversely proportional to the heat flux, suggesting that the HFM should be calibrated using thin transfer standards and high heat fluxes. This calibration would be valid for the low heat flux associated with testing thick specimens only if the output of the thermopile were proportional to the heat flux through it under all test conditions.

Figure 1 shows that this is not always the case even for thin specimens. It gives an example of a calibration performed on a heat flow transducer, using 26-mm thick specimens of three different materials:

low-density glass fiber (LDGF), tested with spacers,
LDGF specimens, tested with wooden frames,
high-density glass fiber (HDGF) specimens.

Figure 1 also indicates that the calibration coefficient obtained for each calibration specimen is different (with probability higher than 95%). There is a difference in the slope of the temperature dependence of the calibration for LDGF and HDGF specimens. In the temperature range 28 to 32°C the cali-

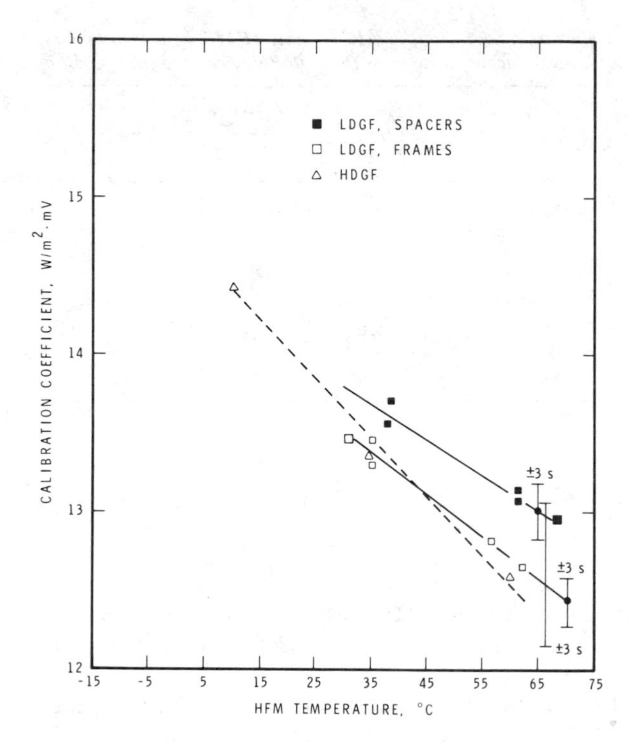

FIG. 1—*Calibration of HFM with 26-mm thick LDGF and HDGF specimens.*

bration coefficient, using LDGF, appears to be the same as that for specimens of HDGF. With extreme temperatures the differences are larger.

As the precision of an HFM calibration using LDGF specimens is as good as that for the high-density standard reference material, it should be possible to obtain the same precision over a larger range of specimen thicknesses or temperatures. This issue has been discussed elsewhere [*8, 9*]. Although calibration of the HFM apparatus in the same range of conditions as will be used in testing identical material (or similar) may constitute a practical solution, a design that will ensure the same calibration coefficient for various materials and testing conditions appears to be a more satisfactory approach.

The response of the heat flow transducer to the properties of the tested specimen may be affected by nonuniformity of the transducer core, which may have low resistance paths through the thermopile junctions. Output will be slightly higher if high conductivity material contacts the surface in the area of the thermopile junctions. This effect can be eliminated by facing the transducer core with a high conductivity layer such as a few metal plates, separated with air gaps, to provide an isothermal surface adjacent to the specimen without increasing the lateral heat flow component at the metering area. Alternatively, the HFM plate can be faced with a layer of low conductivity material such as 1.5 or 2-mm cork. This allows the temperature of the surface adjacent to the specimen to vary, depending on the nature of the contact resistances with the specimen. In both cases the thermal regime at the transducer surface will be always the same, regardless of the properties of the specimen. Two HFM instruments based on these considerations were built at NRCC [*1*].

Other transducer constructions were also tried, for example, dividing the metering field into a number of smaller, separately recorded areas to assess the errors caused by nonuniformity of either the specimen or the transducer itself. In calibrating the apparatus under a given set of testing conditions one has to examine the effect of a specific change in testing conditions on the overall accuracy of thermal resistance testing. This kind of activity is called verification of HFM uncertainties. Different types of transfer standards are required for this purpose.

Transfer Standards for HFM's Error Verification

The following materials were used at DBR/NRCC to establish sets of transfer standards based on uniformity, stability, and handling qualities:

1. Expanded polystyrene with density not less than 18 kg/m^3.
2. Air-filled extruded polystyrene.
3. Medium- and high-density glass fiber insulation (Standard Reference Material No. 1450 from National Bureau of Standard [*10*].
4. Thin (nonconvective) air layers with or without reflective surfaces.
5. Thin layers of glass fiber insulation with a density between 7 and 10 kg/m^3.

A transfer standard is established as follows. A pair of specimens matched in both thermal resistance and density are tested at least nine times in the GHP apparatus. Five of the tests are performed with a mean temperature in the range 20 to 28°C, and four are performed at the extreme ends of the test range, for example, a mean temperature between 0 and 10°C and one between 40 and 50°C. If the standard deviation calculated for these test series is less than 0.5% and a substantial amount of information exists on the same lot (batch) of the material from which the specimens were selected, this pair of specimens will be called transfer standards. If, however, either the standard deviation is higher than 0.5% or there are no other test results on the same lot of the material, the specimens will be called calibrated specimens. They can be used for comparison purposes such as verification of the HFM apparatus errors. However, they are not recommended for calibration of the HFM apparatus. The following sections describe examples of a transfer standard; for other examples see Ref *8*.

Glass Fiber Transfer Standard for Calibration of 300-mm HFM Apparatus

Two specimens, coded 357-154-A and 357-154-B, were prepared from a 1.2 by 2.4 m by 25-mm sheet of glass fiberboard produced by Johns-Manville for the National Bureau of Standards, Washington, D.C., in 1978. Specimens were cut and placed in frames in June 1980.

Density, Specimen A = 56.1 kg/m^3 (3.50 lb/ft^3)
Density, Specimen B = 56.5 kg/m^3 (3.53 lb/ft^3)
Mean = 56.3 kg/m^3 (3.51 lb/ft^3).

Table 3 gives the values of thermal resistance measured in the GHP apparatus, with corresponding values of mean specimen temperature, thickness, and thermal conductivity.

Thickness measured in the GHP apparatus is allowed to differ ±0.2% from the specimen thickness determined outside the apparatus. If it differs by more than 0.2%, the test results are excluded. For a difference of less than 0.1% the measured thermal resistance value is used; for a difference between 0.1% and 0.2% a correction is applied.

The following relation was obtained from a regression analysis of the test results

$$R = 0.8883 - 0.00295T \qquad 0 \leqslant T \leqslant 50°\text{C} \tag{1}$$

where

R = thermal resistance, m^2 K/W and
T = mean specimen temperature, °C.

The mean standard deviation between measured R, recalculated to reference thickness, and the R-values calculated from Eq 1 is s = 0.2%. An error of ±0.4, therefore, can be estimated on a 90% probability level by using two standard deviations.

Use of Expanded Polystyrene for Transfer Standards

Three groups of specimens cut from a special run of expanded polystyrene were examined. Spatial variability in density and in associated thermal conductivity was found to be too large in materials with density less than 18 kg/m^3. Two other series of specimens from materials with a density range of 18 to 24 kg/m^3 (1.1 to 1.5 lb/ft^3) showed satisfactory results. Table 4 shows thermal conductivity values as a function of specimen density measured at a mean temperature of 24°C. Using a linear regression for thermal conductivity as a function of density and comparing this with the measured values, there is a ±1.0% uncertainty range for a probability of 95%, that is, precision com-

TABLE 3—*Thermal resistance of 357-154A/B medium-density glass fiber specimens (GHP tests), specimens dried prior to testing.*[a]

	Measurements			
Date	Mean Temperature, °C	Thermal Resistance, m^2 K/W	Thickness, mm	Thermal Conductivity, W/m K
800716	−0.69	0.8882	25.99	0.02926
800718	12.40	0.8510	25.99	0.03054
800910	23.96	0.8184	25.97	0.03174
800912	22.21	0.8205	25.97	0.03165
800917	25.97	0.8114	25.97	0.03200
800918	28.04	0.8045	25.97	0.03228
800922	35.92	0.7828	25.97	0.03318
800929	47.44	0.7509	25.97	0.03458
801001	0.19	0.8933	25.97	0.02907
801003	12.44	0.8514	25.97	0.03050
801006	23.28	0.8165	25.97	0.03181
801027	20.14	0.8282	25.99	0.03138
801028	23.72	0.8185	25.99	0.03176

[a]Reference thickness, $L = 25.97$ mm.

TABLE 4—*Thermal conductivity and density of polystyrene specimens tested on 600-mm square heat flow meter apparatus at 24°C.*

Specimen Code	Density, ρ kg/m^3	Measured λ, W/m K
1	21.6	0.0375
2	18.1	0.0405
3	22.7	0.0365
4	24.0	0.0352
5	19.2	0.0399
6	22.6	0.0370
7	23.0	0.0365
8	22.3	0.0373
9	22.2	0.0371

parable with that obtained on high-density glass fiber transfer standards. Temperature dependence for three sets of transfer standards prepared for the same lot of material is shown in Table 5. The agreement is good.

Use of Layered, Low-Density Glass Fiber Specimens as Transfer Standards

The following technique was used at DBR/NRCC to fabricate low-density transfer standards ranging in thickness from 50 to 200 mm. Using a light

TABLE 5—*Thermal resistance as a function of temperature for three sets of transfer standards and calibrated specimens prepared from the same batch of 18.4 kg/m³ density expanded polystyrene.*

Specimen Code	Temperature Dependence, R (m^2 K/W) versus T_m (°C)
365-146	$R = 2.792$ (1 to 0.00330 T_m)
372-173	$R = 2.773$ (1 to 0.00318 T_m)
372-2	$R = 1.343$ (1 to 0.00335 T_m)

table [*11*], 30 specimens were selected whose density varied between 7.2 and 10.0 kg/m³. To reduce the number of tests required to establish their thermal characteristics two specimens separated by one paper layer (septum) to intercept radiative heat transfer were tested. Nine pairs were finally selected [*8*].

Thermal conductivity tests on glass fiber multilayer specimens with and without paper septa were useful in examining the effect of specimen thickness [*12,13*] on apparent thermal conductivity and HFM apparatus errors related to increasing specimen thickness [*8,9*]. Although thick, uniform, calibrated specimens are developed in a manner identical to that used for transfer standards (see description of the glass fiber transfer standard in the previous section), the development of layered, calibrated specimens is far more difficult. The thermal resistance of these specimens is determined for standard conditions (24°C mean) only. To determine such values DBR/NRCC uses GHP and HFM measurements as well as computer calculations.

Proficiency Testing

In contrast to transfer standards, which are loaned by DBR/NRCC together with certificates stating both the mean value and the uncertainty in thermal resistance over a given temperature range, calibrated specimens for proficiency testing are made available without any information on their thermal resistance. Each laboratory must decide on a testing procedure for determining the thermal resistance of the calibrated specimens; it knows only that different results may be obtained with different testing conditions.

There are two kinds of proficiency tests: those performed at standard conditions of mean temperature 24 ± 1°C and temperature difference of 22 ± 2°C; and those performed over a range of mean temperatures between 0 and 50°C. Although some requirements for the two cases will be different, the same specimens are used. Those for proficiency testing will include homogeneous transfer standards and the layered, calibrated specimens specially constructed for the purpose of proficiency testing.

Use of Layered Specimens

At DBR/NRCC the following were built and developed as calibrated, layered specimens for proficiency testing:

1. 180 to 190-mm thick, uniform, layered polystyrene (type 2),
2. 150 to 180-mm thick, uniform, layered, low-density glass fiber,
3. 90 to 120-mm thick, cellular plastic specimens covered on one surface with medium conductive, uneven, metal reinforced bonding cement surface.

Table 6 shows thermal resistance of one pair of layered specimens as tested on HFM and GHP apparatus.

Proposed Requirements for Standard Test Conditions

1. The laboratory should perform, in sequence, a minimum of eight tests on one pair of transfer standards: Specimen A alone; Specimen B alone; Specimens A and B together, with a paper septum between. The mean thermal conductivity should be within 2.0% of the mean value determined in the GHP apparatus at DBR/NRCC, and the standard deviation of thermal resistance determined in this test series should not exceed 1.0% (that is, twice that allowed for transfer standards at DBR/NRCC).

2. Three coded, layered, calibrated specimens should be tested. The mean test results should not differ more than 3% from the mean determined at NRCC, and none of the separate test results should differ more than 5% from the mean value determined by NRCC.

Proposed Requirements for Testing over a Temperature Range

1. The laboratory should perform a minimum of nine tests on each pair of transfer standards, five with mean temperature in the range 20 to 28°C, and

TABLE 6—*Apparent thermal resistance of 90-mm thick polystyrene specimen covered on one side with bonding cement reinforced by expanded metal lath to increase lateral conduction of surface layer.*

Equipment Type	Test Code	Heat Flux Measured on Side Contacting		Placement of Stucco
		Stucco	Polystyrene	
GHP	380-30		100.0	cold side
GHP	380-29	94.3		hot side
HFM	380-24	95.3	99.5	hot side
HFM	380-14	93.2	98.1	hot side
HFM	380-25	96.2	100.1	cold side
HFM	380-15	96.0	101.1	cold side
HFM mean value	...	95.2	100.3	...

four at the extreme ends of the test range, for example, between 0 and 10°C and between 40 and 50°C. Thermal conductivity should be within the following limits:

(*a*) Mean of three or more tests at 24 ± 2°C should be within 2.0% of the NRCC determined value.

(*b*) Mean of two or more tests at other temperature should not differ by more than 2.5%.

2. Three layered, calibrated specimens should be tested at 24 ± 1°C. The requirements are the same as for point 2, Standard Test Conditions.

3. Lateral heat flow tests with three different ambient air temperatures should also be performed [*8*].

Closing Remarks

The transfer standards and calibrated specimens must be returned to DBR/NRCC. In addition, the laboratory applying for accreditation should develop its own reference specimen. It may be one specimen or a pair of identical specimens (material may be supplied by NRCC). The verification tests will be performed at DBR/NRCC and the specimens then returned to the accredited laboratory. Control charts must be established by that laboratory and adjusted during subsequent testing of the reference specimen [*9*]. The reference specimen must be retested by the accredited laboratory periodically and not more than 30 days before the test to be reported.

Cooperative projects performed by DBR/NRCC and by testing and industrial laboratories have proved that HFM apparatuses may be recalibrated quite easily to meet the proposed requirements.

Acknowledgment

The authors wish to thank J. G. Theriault, R. G. Marchand and Nicole Normandin of the Thermal Insulation Laboratory, DBR/NRCC, for their contribution to the development of testing facilities transfer standards and for making the measurements. They also wish to express their gratitude to C. M. Pelanne for detailed discussion of HFM apparatus calibration.

This paper is a contribution from the Division of Building Research, National Research Council of Canada, and is published with the approval of the Director of the Division.

References

[*1*] Bomberg, M. and Solvason, K. R. in *Thermal Insulations, Materials, and Systems for Energy Conservation in the '80s, ASTM STP 789*, American Society for Testing and Materials, Philadelphia, 1983, pp. 277–292.

[2] Bomberg, M. and Solvason, K. R. in *Proceedings*, 17th International Thermal Conductivity Conference, Plenum Press, New York, pp. 393–410.
[3] Woodside, W., *Thermal Conductivity Measurements, ASTM STP 217*, American Society for Testing and Materials, Philadelphia, 1957, pp. 49-62.
[4] Woodside, W. and Wilson, A. G., *Thermal Conductivity Measurements, ASTM STP 217*, American Society for Testing and Materials, Philadelphia, 1957, pp. 32-46.
[5] Brown, W. C. and Shirtliffe, C. J. in *Proceedings*, 14th Thermal Conductivity Conference, Plenum Press, New York, 1975, pp. 539-542.
[6] Degenne, M., Klarsfeld, S., and Barthe, M. P. in *Thermal Transmission Measurements of Insulation, ASTM STP 660*, American Society for Testing and Materials, Philadelphia, 1978, pp. 130–144.
[7] Zabawsky, Z., *ISA Test Measurements Symposium*, Vol. 5, Paper No. 68-520, 1968, pp. 1-6.
[8] Bomberg, M., Pelanne, C. M., and Newton, W., "Analysis of Uncertainties in Calibration of a Heat Flow Meter Apparatus," *Thermal Conductivity*, Vol. 18, Plenum Press, New York.
[9] Newton, W., Pelanne, C. M., and Bomberg, M., "Calibration of Heat Flow Meter Apparatus Used for Quality Control of Low Density Mineral Fiber Insulations," *Thermal Conductivity*, Vol. 18, Plenum Press, New York.
[10] Siu, M. C. I. in *Thermal Insulation Performance, ASTM STP 718*, American Society for Testing and Materials, Philadelphia, 1980, pp. 343-360.
[11] Pelanne, C. M. in *Thermal Transmission Measurements of Insulation, ASTM STP 660*, American Society for Testing and Materials, Philadelphia, 1978, pp. 263-280.
[12] Pelanne, C. M. in *Thermal Insulation Performance, ASTM STP 718*, American Society for Testing and Materials, Philadelphia, 1980, pp. 322-334.
[13] Alberts, M. A. and Pelanne, C. M. in *Proceedings*, 17th International Thermal Conductivity Conference, Plenum Press, New York, 1983, pp. 471-481.

David J. McCaa[1]

Use of a Heat Flow Meter Apparatus for Testing Loose-Fill Materials Prepared for Laboratory Testing

REFERENCE: McCaa, D. J., **"Use of a Heat Flow Meter Apparatus for Testing Loose-Fill Materials Prepared for Laboratory Testing,"** *Guarded Hot Plate and Heat Flow Meter Methodology ASTM STP 879*, C. J. Shirtliffe, and R. P. Tye, Eds., American Society for Testing and Materials, Philadelphia, 1985, pp. 154–160.

ABSTRACT: The use of loose-fill type attic insulation is increasing, and the requirement for manufacturers to test their products has also increased along with production. This paper discusses some of the critical parameters involved in testing loose-fill materials using a heat flow meter apparatus. It is shown, in studies conducted to characterize a material, that many of the sample preparation variables must be fixed.

KEY WORDS: glass fiber, building Insulation, thermal conductivity, heat flow meter apparatus, thickness effect, sample preparation, loose fill testing

Over the past three years the thermal laboratory of the CertainTeed Corporation has been deeply involved in the testing of loose-fill materials. Not only our own but also all commercially available loose-fill products have been tested. The testing of loose fills with a heat flow meter (HFM) apparatus is currently covered by ASTM Standard Recommended Practice for Determination of the Thermal Resistance of Low-Density Fibrous Loose Fill-Type Building Insulation (C 687-71).

This practice is now undergoing revision in ASTM Committee C-16 on Thermal Insulation to reflect, in part, what the industry has learned about testing loose fills. This is an extremely complex subject area, and, in this paper, we will limit its discussion to the HFM apparatus and some of the areas of concern directly related to the preparation of loose-fill samples.

One commercially available instrument which is typical of those employed for testing loose fills is shown in Fig. 1. While many instruments exist in

[1]Supervisor, Thermal Testing Laboratory, CertainTeed Corporation, Blue Bell, PA 19422.

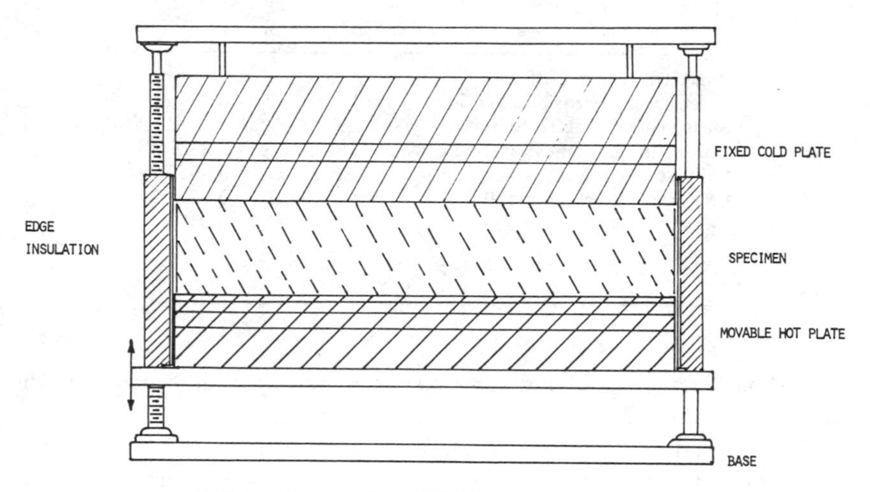

FIG. 1—*Cross section of HFM apparatus test section.*

which the heat flow transducer is in a vertical configuration, the methods of sample insertion and preparation are far more complex than those required for a horizontal apparatus such as the one shown. This particular apparatus allows for plate separations of up to 250 mm (10 in.) and has the actual transducer located on the lower (hot) plate with the cold plate at the top. The particulars of the plate construction have been previously described [*1*]. Note the edge insulation present at the sides of the sample area.

In typical use (up to 150 to 178 mm of sample thickness) the edge insulation completely fills the area between the hot and cold plates. However, due to space restrictions, when the sample thickness exceeds 200 mm there can be an air gap next to the cold plates.

The possible effect of the air gap, if any, is difficult to determine. There currently exist calibration standards available from the National Bureau of Standards (NBS) for low density batts which are valid up to 152 mm (6 in.) thick. If one argues that any increase in edge losses due to air gap would show up in higher values of the measured apparent thermal conductivity, λ, it must be realized that for loose-fill material, blown at a constant weight/metre square (weight/foot square), at thickness over 200 mm (8 in.) the samples are at a very low density and the curves of λ versus density are very steep, making error detection difficult.

In an attempt to evaluate this potential problem the NBS low density calibration standards were tested at a 228 mm (9 in.) test thickness by stacking three 77 mm (3 in.) specimens. The results are shown in Table 1. The full thickness calibration factor found at 228 mm is nearly the same as the one we must use at 150 mm. Since both 150 and 228 mm are well beyond the representative thickness of 80 mm for this low density material [*2*], the calibration

TABLE 1— *Low density calibration data.*

Projected value of NBS low density standards at 228 mm	$\lambda = 0.339$
Measured value at 228 mm using the 25 mm calibration constant	$\lambda = 0.352$
Calibration constant at 228 mm	$C_o = 3.96$
Calibration constant at 150 mm	$C_o = 3.97$
Calibration constant at 25 mm	$C_o = 4.05$

$$\text{Where } \lambda = C_o \left(\frac{Q \cdot \Delta X}{\Delta T}\right)$$

factors should agree at 150 and 228 mm—if no new apparatus errors are introduced at 228 mm.

From this result, it was concluded that the air gap in the edge insulation at thicknesses over 200 mm does not significantly effect the measured heat flow.

For all these testing programs loose-fill samples were prepared using a pnuematic blowing machine, and specimens were blown into sample frames constructed of 6 mm (1/4 in.) plywood sides, with fiber glass window screen on the bottom of the frame. The outside of the frame is 600 by 600 mm (24 by 24 in.) so that it fills the sample chamber of the apparatus, as seen in Fig. 2. The height of the frame side is selected to be at least 10 mm less than the minimum test thickness. Whenever specimens are moved in the Laboratory, a cardboard support is used under the frame.

In any loose-fill test program in which the objective is to study material variability or properties, it is extremely important (if not mandatory) that the sample preparation techniques be fixed. Consider Fig. 3 where k-density curves for two different types of loose-fill materials are shown. For both materials only one lot of material was used to avoid sample variations caused by production variables. All the differences observed in Fig. 3 were caused by

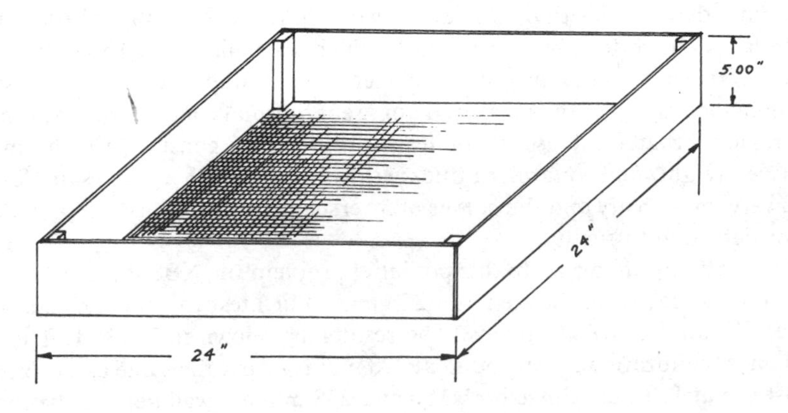

FIG. 2—*Loose-fill sample frame (1 in. = 25.4 mm).*

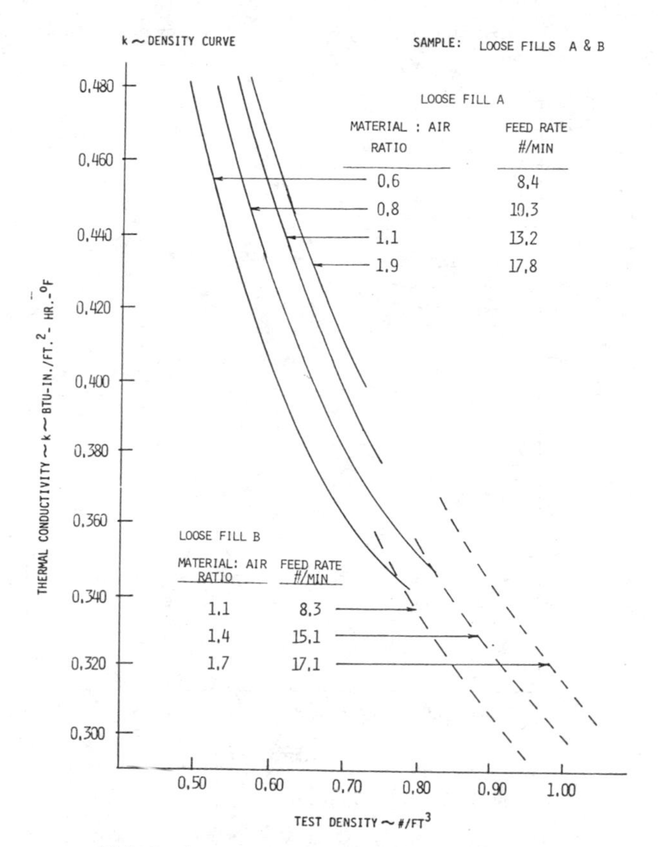

FIG. 3—*k versus density for loose-fill insulations.*

changing the blowing machine settings. A constant 50 m (150 ft) of 75-mm (3-in.) corrugated hose was used, and all samples were prepared by one operator. It should be obvious from these results that using a HFM apparatus for a quality control or certification program can quickly lead to erroneous conclusions if sample preparation techniques are not fixed.

Typically k-density curves such as these are generated by blowing the specimen to a lighter density and then compressing the specimen in the apparatus by reducing the plate separation. In order to see the validity of this procedure a series of samples were prepared with blown thicknesses from 50 to 178 mm. These samples were all blown at constant machine conditions, and only the hose angle with respect to the sample frame was changed to vary the installed density and thickness. These were then tested at only one thickness, with a constant 15% compression. The resultant data points are shown in Fig. 4 along with a normal k-density curve for the same material generated by compressing a single sample from 7.0 to 5.0 in. The differences seen between the

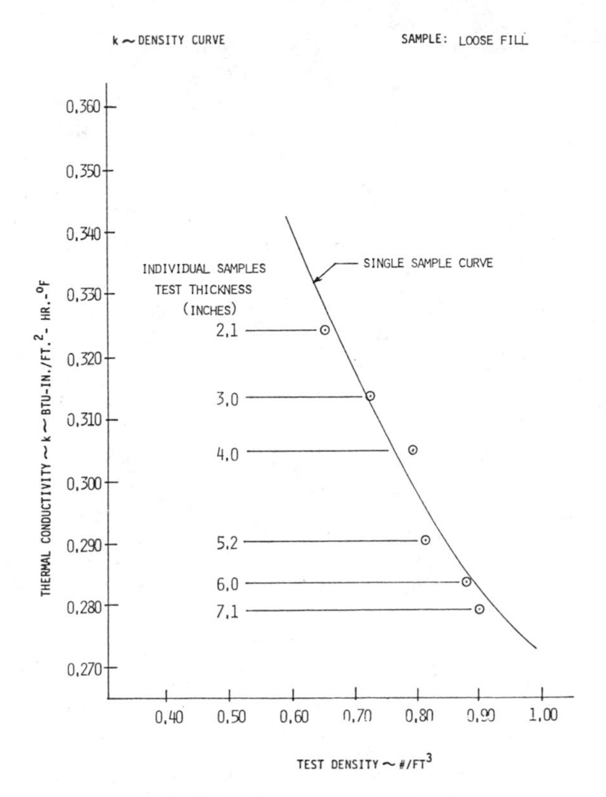

FIG. 4—*k versus test thickness.*

individual points and the single sample curve are well within the experimental error.

Once this point has been accepted the number of specimens which have to be tested is reduced. Studies at NRC Canada [*3*] have confirmed our conclusion that compression of samples blown thicker than the final test thickness is acceptable if not carried to extremes. For a 50 to 60 mm overblow on a 150-mm specimen, they found only a 2% change in k-value between the overblown sample and a specimen with only 15 to 25 mm overblow. In fact, due to the nonhomogeneous nature of some loose-fill materials, it is necessary to achieve some compression (in the range of 10 to 15% minimum) of the top surface of the specimens in order to assure good plate contact in the HFM apparatus.

If sample material conditioning is fixed in the blowing machine it can be shown that specimens of varying overblow will fall on the same k-density curve, provided that they are blown to the same grams/metre square (pounds/foot square) of material in the test area. Consider the curves shown in Fig. 5. Here specimens were blown to constant weight/metre square (weight/foot square) with up to 60% greater loft than the test thickness from

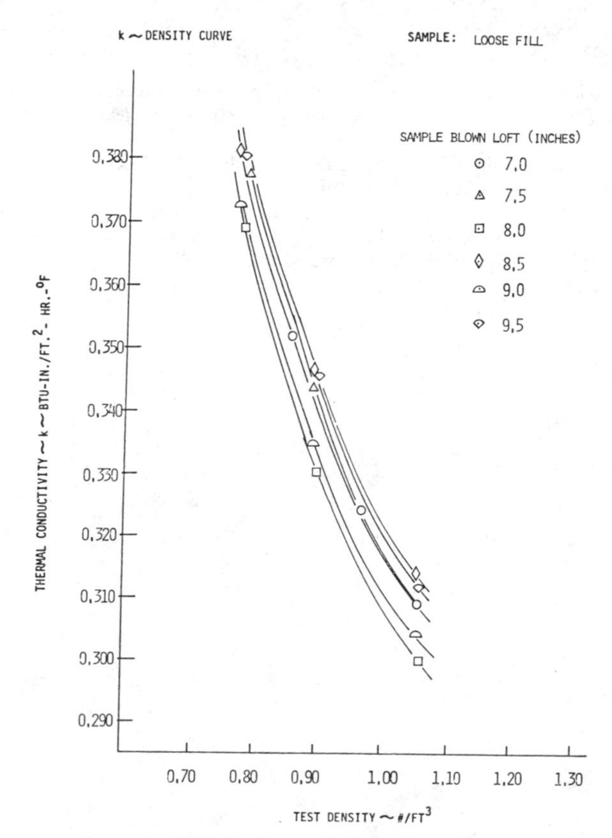

FIG. 5—*k versus blown thickness.*

one sample lot. All specimens were then tested in the same HFM apparatus at the same three test thicknesses, (178, 165, and 152 mm).

Considering the fact that six individual specimens are involved, the agreement is satisfactory, with all six exceeding the minimum bag label for this product. Note that there is no systematic shift in the k-density curves as one goes from 177 to 237 mm (7 to 9.5 in.) thickness. Thus a quality control test could be made on any of the specimens, irrespective of the loft.

In conclusion it has been shown that an HFM apparatus can be a useful tool in evaluating and characterizing loose-fill products. However, in order to achieve meaningful and consistant results, one must be very careful to standardize the parameters involved in the sample preparation.

Acknowledgments

The author would like to thank J. Carré, for his assistance in the preparing of the test specimens and R. Palat for her assistance in the testing of the specimens.

References

[*1*] Degenne, M., Klarsfeld, S., and Barthe, M. P. in *Thermal Transmission Measurements of Insulation, ASTM STP 660*, American Society for Testing and Materials, Philadelphia, 1978, p. 130.

[*2*] "Labelling and Advertising of Home Insulation," Federal Trade Commission, Bureau of Consumer Protection, Final Staff Report, July 1978.

[*3*] Bomberg, M. and Solvason, K. R., "How to Ensure Good Thermal Performance of Blown Mineral Fiber Insulation in Horizontal and Vertical Installations," NRCC Report No. 167, National Research Council of Canada, Ottawa, Canada, 1980.

Karl G. Coumou[1]

A Review of Current Commercially Available Guarded Hot Plate and Heat Flow Meter Apparatus and Instrumentation

REFERENCE: Coumou, K. G., **"A Review of Current Commercially Available Guarded Hot Plate and Heat Flow Meter Apparatus and Instrumentation,"** *Guarded Hot Plate and Heat Flow Meter Methodology, ASTM STP 879*, C. J. Shirtliffe and R. P. Tye, Eds., American Society for Testing and Materials, Philadelphia, 1985, pp. 161-170.

ABSTRACT: Equipment for the measurement of thermal performance is commercially available from a number of sources. Specifications of the various apparatus are compared, and unique features of each model are described briefly in the text. Future requirements for commercially produced thermal testing equipment are discussed.

KEY WORDS: guarded hot plate, heat flow meter, computer aided testing, thermal conductivity, commercial test apparatus

Thermal conductivity instruments are commercially available from a number of companies. For the purpose of this review requests for technical information were sent to the following organizations:

Custom Scientific Instruments, Inc., Whippany, NJ, USA
Hauser Laboratories, Boulder, CO, USA
Sparrell Engineering, Damariscotta, ME, USA
Dynatech R/D Company, Cambridge, MA, USA
Foundation Electronic Instruments, Ottawa, Ont., Canada
Anacon, Inc., Burlington, MA, USA
Karl Weiss, Giessen, W. Germany
Yarsley Laboratories, Chessington, Surrey, UK

All but the last three firms have responded to the inquiry. Some information on the Anacon unit was obtained from one-year-old sales literature. No sales

[1]Vice President, Dynatech R/D Company, Cambridge, MA 02139.

documentation was available on the Weiss and Yarsley equipment, both believed to be heat flow meter (HFM) apparatus. Consequently, the current survey includes equipment made in North America only.

Review

Table 1 compares the specifications of seven HFM instruments ranging in plate size from 20 to 91 cm.

The Hauser unit (Fig. 1) employs duplex HFMs on both the upper and lower surfaces. The duplex HFM is 46 cm^2 of which the center 30 by 30 cm^2 area can be measured separately. Dried air can be blown into the space surrounding the sample using a built-in air pump and rechargeable cartridge with dessicant. Hauser reports that edge effects and sample distribution effects can be detected with the aid of duplex HFMs.

The Foundation Model HFMA-101 (Fig. 2) offers optional operation at plate temperatures down to −30°C. Dried air can be blown into the sealed test chamber from an external supply. The edges of the specimens are exposed to surfaces that are temperature controlled to avoid lateral heat exchange.

The Dynatech R-Matic is available with a built-in air conditioning system to dehumidify the air surrounding the specimen, and control the air temperature at the mean specimen temperature. The R-Matic hot surface plate is constructed as a bi-guarded hot plate allowing this HFM instrument to be operated in the absolute (guarded hot plate) as well as the relative (HFM) mode.

The Sparrell Engineering Model TC-100 utilizes two HFMs, one above the specimen and one below. One purpose of the two HFMs, according to Sparrell, is to account for edge heat transfer during the measurement of apparent thermal conductivity, especially important when investigating the effect of thickness in low density insulations. Sparrell reports measuring only about half the thickness effect from those using a single HFM.

Both the Anacon Model 88 and Dynatech k-Matic feature a built-in analog calculator with digital display of the k-factor. The Dynatech Rapid-k (Fig. 3) is offered with software for computer-aided testing using a Hewlett Packard HP-85 microcomputer. Electrical outputs from the test chamber (temperatures, HFM signal, specimen thickness) are read periodically by a data scanner. The computer is programmed to recognize the onset of thermal equilibrium; it then calculates and records the test results and advances the heater controllers to the next set point as instructed by the operator via the key board. Computer aided operation of a HFM apparatus is particularly useful when measuring thermal conductivity over a range of temperatures.

Table 2 summarizes the specifications of seven commercially available guarded hot plate (GHP) instruments. Plate sizes range from 20 cm round to 61 cm square. Materials can be tested from about −160°C, using liquid ni-

TABLE 1—*Comparison of commercially available HFM apparatus.*

Manufacturer Model No.	Anacon, 88	Dynatech, k-Matic	Dynatech, Rapid-k	Dynatech, R-Matic	Foundation, HFMA-101	Hauser Labs, Duplex HFM	Sparrell Eng. Co., TC-100
Plate dimensions, cm by cm[a]	20 by 20 or 30 by 30	30 by 30	30 by 30	61 by 61	61 by 61	91 by 91	46 by 46
Metering area, cm by cm	N/A[b]	10 by 10	10 by 10	25 by 25	25 by 25	30 by 30 or 46 by 46	20 by 20
Maximum specimen thickness, cm	5	10	10	24	27	23	20
Heat flow direction	down	down	up/down	up	up	up/down	down
Heat flux transducer at hot surface	yes	no	optional	yes	yes	yes	yes
Heat flux transducer at cold surface	no	yes	yes	optional	optional	yes	yes
Mean specimen temperature, °C	24	10 to 80	−12 to 200	10 to 40	−30 to 110[c]	−18 to 150[c]	10 to 32
Specimen environment	ambient air	ambient air	ambient air	dried and temperature controlled air	dried air	dried air	ambient air
Test method	N/A	ASTM C 518	ASTM C 518	ASTM C 518[d]	ASTM C 518	ASTM C 518 and C 687	ASTM C 518

NOTE: All specifications are taken from manufacturer's literature.

[a] All plates are square.

[b] NA = not available.

[c] Plate temperature.

[d] Also usable as bi-guarded hot plate.

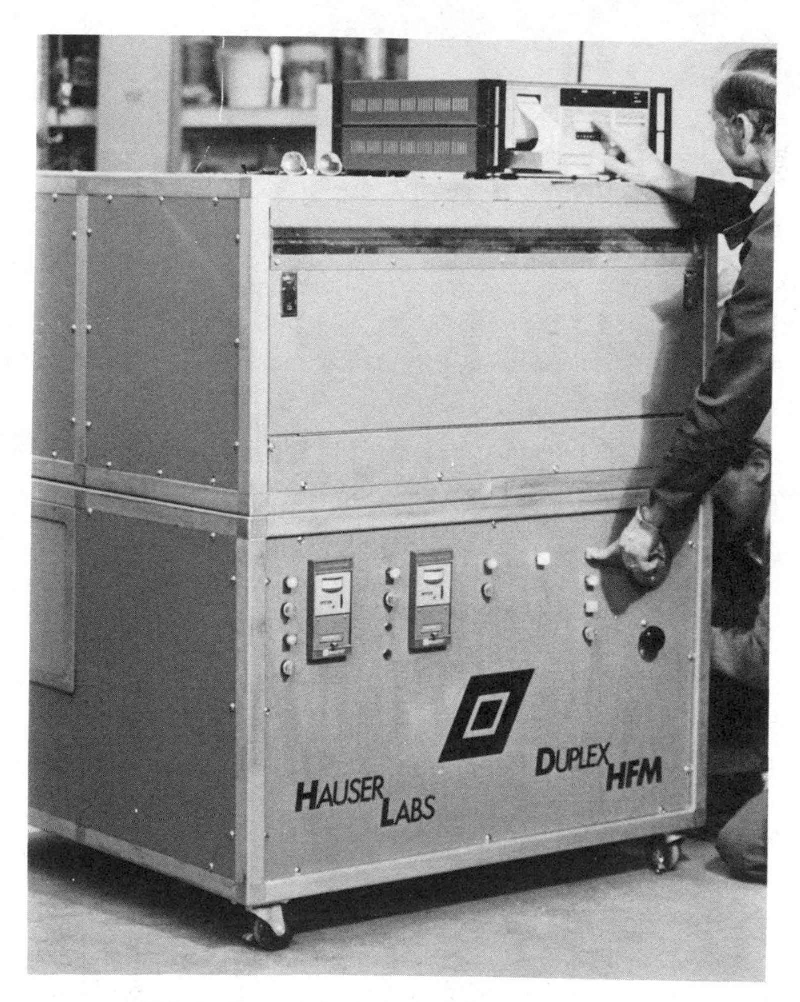

FIG. 1—*Hauser Labs duplex heat flow meter apparatus.*

trogen for cooling of the heat sink, to 950°C mean sample temperature. The sample environment can be ambient air, vacuum, or gas.

The Custom Scientific Model CS-76 (Fig. 4) is available with either a 23-cm-diameter alundum heater without metal surface plates, or a 30-cm square heater having metallic surface plates. The Dynatech Model TCFGM uses 20-cm-diameter heaters made of heater wire or foil embedded in, depending on allowable operating temperatures, silicone rubber, bonded mica, or aluminum oxide, all with metal surface plates. The unit is available for testing in a vacuum or controlled gas environment.

The Sparrell Engineering Model GP-500 (Fig. 5) uses silicone rubber encapsulated heating elements with metal surface plates. The Foundation GHP

FIG. 2—*Foundation Instruments 61-cm heat flow meter apparatus.*

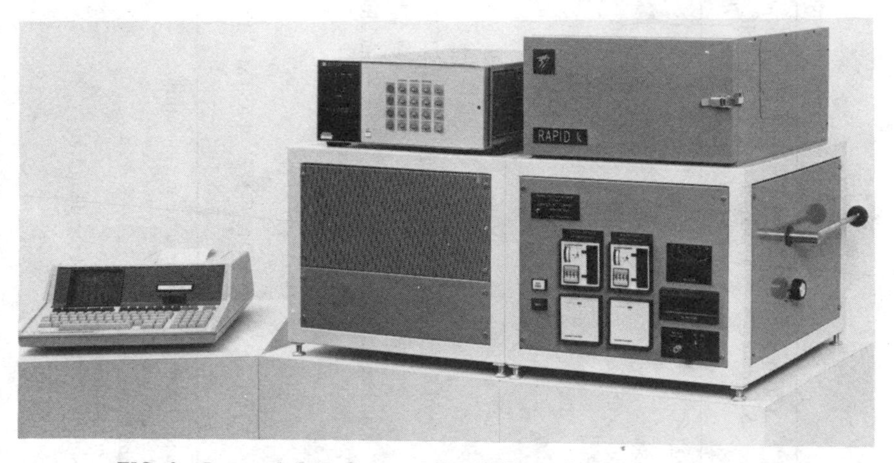

FIG. 3—*Dynatech R/D Company Rapid-k heat flow meter apparatus.*

is provided with a system for rotating the entire test chamber from a vertical to horizontal specimen position. The instrument utilizes a double guard around the hot plate.

The Dynatech Model HTLM is a bi-guarded hot plate for repetitive testing of materials at temperatures to 650°C. The unit allows specimens to be ex-

TABLE 2—*Comparison of commercially available GHP apparatus.*

Manufacturer Model No.	Custom Scientific, CS-76	Dynatech, TCFGM	Dynatech, TCFG-R4-6	Dynatech, TCFG-R4-4	Dynatech, HTLM	Foundation	Sparrell Eng. Co., GP-500
Plate dimensions,[a] cm by cm	23 ϕ or 30 by 30	20 ϕ	61 by 61	46 by 46	30 by 30	61 by 61	20 ϕ or 30 by 30
Metering area, cm by cm	15 ϕ of 20 by 20	10 ϕ	30 by 30	25 by 25	15 by 15	30 by 30	10 ϕ or 15 by 15
Maximum specimen thickness, cm	5	5	15	10	4[b]	36[c]	5
Maximum hot plate temperature, °C	540 or 320	1000	230	230	650	room temperature	230
Minimum cold plate temperature, °C	ELCT[d]	−180 with LN_2	ELCT[d]	−180 with LN_2	100	ELCT[d]	ELCT[d]
Specimen environment	ambient air	vacuum, gas ambient air	ambient air	vacuum, gas ambient air	ambient air	ambient air	ambient air
Number of specimens required	2	2	2	2	1	2	2
Test method	ASTM C 177	ASTM C 177	ASTM C 177	ASTM C 177	bi-guarded hot plate	ASTM C 177	ASTM C 177

NOTE—All specifications are taken from manufacturer's literature.

[a] ϕ = round plate. All others are square.

[b] Minimum specimen thickness is also 4 cm.

[c] Maximum available test chamber opening.

[d] ELCT = externally supplied liquid coolant temperature.

FIG. 4—*Custom Scientific Instruments guarded hot plate apparatus.*

changed rapidly at any temperature level without exposing the heater surface plates to ambient air.

All GHP instruments listed in Table 2, except for the Dynatech Model HTLM, are designed to conform to ASTM Test Method for Steady-State Thermal Transmission Properties by Means of the Guarded Hot Plate (C 177-76). Three manufacturers, Custom Scientific, Sparrell Engineering, and Dynatech, also report the building of special GHP instruments for the testing of textile fabrics according to ASTM Test Method for Thermal Transmittance of Textile Fabric and Batting Between Guarded Hot-Plate and Cool Atmosphere (D 1518-77). These instruments are designed as bi-guarded hot plates with the plate surface controlled at 33°C to simulate the temperature of the human skin. A fabric specimen is placed on the bi-guarded hot plate. Heat is dissipated from the specimen to the surrounding air by conduction

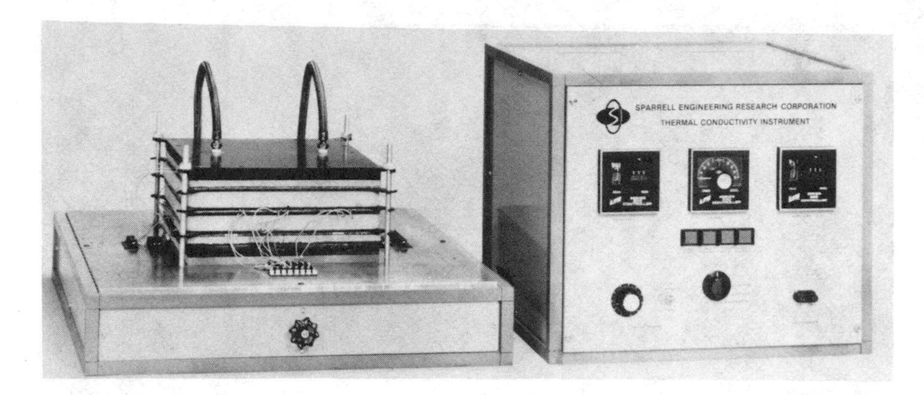

FIG. 5—*Sparrell Engineer Research Corporation 30-cm guarded hot plate apparatus.*

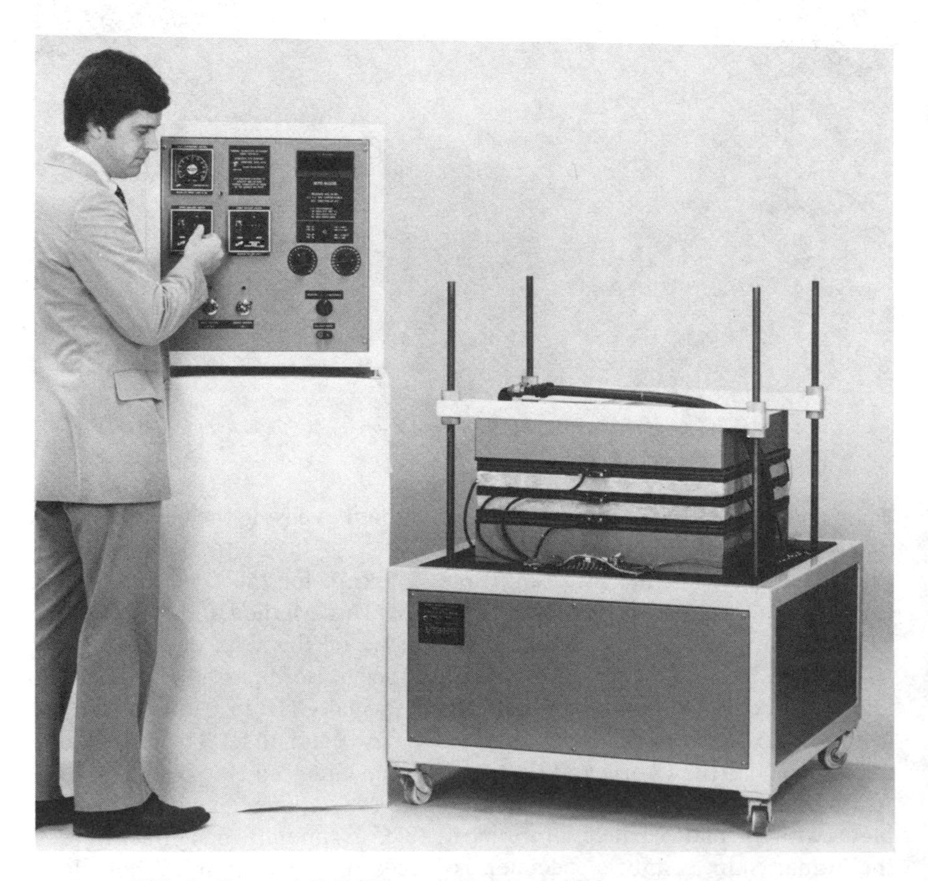

FIG. 6—*Dynatech R/D Company 61-cm guarded hot plate apparatus.*

and convection. The overall thermal resistance of the material between the hot plate and the air above the specimen is, in the United States, usually expressed in terms of clo, where 1 clo = 1.137°F ft^2 h/Btu. Dynatech has recently constructed a special version of the ASTM D 1518 GHP apparatus, featuring continuous moisture vaporization at the surface of the plate to simulate perspiration from the skin.

No prices are listed for any of the instruments covered by this survey. Many of the models discussed are available with a variety of options and custom-modifications making simple price comparisons difficult and meaningless.

Design Considerations for Commercial Apparatus

The following comments made by the author, who is a manufacturer of commercial test apparatus, are not necessarily shared by others in the business.

In designing equipment with the widest possible appeal a manufacturer should consider the following, sometimes conflicting, requirements:

1. Accuracy and precision should be consistent with generally accepted industry standards.
2. Equipment should be relatively simple to operate permitting use by inexperienced personnel-operator judgment should be kept to a minimum.
3. Specifications (operating temperatures, specimen sizes, specimen environment, etc.) should include most of the market requirements.
4. Design should be versatile to meet special demands, and flexible to allow custom-modifications as required.
5. Packaging of the product should be both functional and attractive.
6. Equipment should be reliable and permit easy serviceability in the event of a breakdown.
7. Cost should be competitive.

Many of these requirements are not of consideration to an expert designing and building equipment for his or her own use. For example, use of sheathed thermocouples improves reliability at high temperatures, but less expensive, small diameter, bare wire thermocouples may permit more accurate temperature measurement by being less of a disturbance in the specimen. Here, reliability is gained at the expense of accuracy and cost. Temperature measurement using a digital thermometer is convenient but less accurate than using a digital millivolt meter with 1 μV precision, thus trading convenience for accuracy. Providing fixed, manual power input to a GHP main heater yields accurate test results and is less expensive than using a temperature controller with programable d-c power supply. With the former system there is, however, no control of the final hot plate temperature, and the operator may have to make occasional power adjustments. Here, the tradeoff is primarily between convenience and cost.

Future Requirements for Commercial Test Apparatus

Following are some of the requirements that are expected to be demanded increasingly of commercially available test equipment:

1. Measurement of thermal conductivity for quality assurance, especially in the thermal insulation industry, should generally be faster and with better accuracy and precision than available at present.
2. There is a need for rapid measurement of thermal conductivity at elevated temperatures (to 600°C), using either a HFM or absolute measurement technique.
3. There is a need for commercially available test equipment to measure accurately materials of very low thermal conductivity (λ below 0.01 W/mK), often requiring a controlled specimen environment (vacuum or gas).
4. Measurement of thermal conductivity for quality assurance in the thermal insulation industry should be carried out in a specimen environment controlled for humidity and ambient temperature.
5. There will be increasing demand for computer aided operation of thermal conductivity test equipment.

Many of these requirements have been developed already and are available at this time. An apparatus for rapid measurement at elevated temperatures is now produced by Dynatech. To measure conductivities down to 0.004 W/mK, Dynatech has produced a 46-cm-square GHP which can be placed inside a 1-m-diameter vacuum chamber. Control of the specimen environment in a HFM apparatus, to prevent moisture condensation and edge heat transfer, has been accomplished in several different ways as discussed briefly in the Review section.

One area of concern with some HFM apparatus has been the slight shift in heat flux transducer surface temperature on the side facing the specimen, when testing materials of widely varying R-values. This shift results from changes in the ΔT through the heat flux transducer coupled with the fact that the temperature on the opposite side of the transducer is fixed by the heater controller. Two approaches have been suggested to reduce the size of the problem. A thinner transducer would have a smaller ΔT. With computer-aided testing the set point of the temperature controller can be adjusted automatically to produce the same heat flux transducer surface temperature for each test.

Yukiharu Miyake[1] *and Kazuo Eguchi*[2]

Development of a Large Area Heat Flow Meter

REFERENCE: Miyake, Y. and Eguchi, K., **"Development of a Large Area Heat Flow Meter,"** *Guarded Hot Plate and Heat Flow Meter Methodology, ASTM STP 879*, C. J. Shirtliffe and R. P. Tye, Eds., American Society for Testing and Materials, Philadelphia, 1985, pp. 171–179.

ABSTRACT: A large area heat flow meter (HFM) is required to measure heat flow from whole area of building walls. Therefore, the cost of the HFM must be as low as possible.

Under such background we developed a large area HFM by means of photoetching and electroplating techniques. The new large area HFM can be also used as a small area HFM by cutting it into small units. Further, we developed a new calibration method which enabled us to calibrate HFM, by both absolute and relative methods at the same time with a single apparatus due to capability of cross checking. The difference of the calibrated values by both methods was not greater than ±1%. For practical use of a large area HFM, calibration of several HFMs must be carried out at the same time; we also described the method in this paper.

KEY WORDS: heat flow meter, calibration, insulating materials

Recently, a large number of heat flow meters (HFMs) have been used to measure the thermal transmission properties of insulating materials and thermal behavior of whole building components under *in situ* field conditions. Since a large number commercial HFMs are only available in 10 by 10 cm sizes, they are unsuitable to measure accurately the previously mentioned properties without some modification [*1*].

The measurement of a limited small area does not usually indicate the overall properties of a whole building component, and also the placing of an HFM having a relatively small area on a surface leads naturally to some error due to thermal disturbance. Furthermore, if we attach the meters to the whole area, the cost might be considerably increased.

[1]Products and technical director, EKO Instruments Trading Co., Ltd., 21-8 Hatagaya 1-chome, Shibuya-ku, Tokyo 151, Japan.

[2]Principal research director, Building Research Institute, Ministry of Construction, Tachihara-1, Oho-machi, Tsukuba-gun, Ibaragi-Pref., Japan.

As an example, a large area HFM will be required to evaluate the heat losses through the side walls of a guarded hot box. The walls themselves can be constructed using HFMs. A large area HFM which is thermally uniform in the area of 30 by 30 cm^2 is also required for a thermal conductivity measurement according to the HFM method [for example, ASTM Test Method for Steady-State Thermal Transmission Properties by Means of the Heat Flow Meter (C 518-76)].

Furthermore, if a temperature sensor and a plane heater are also included in an HFM, the total thermal measurement can be carried out more accurately.

Based upon these considerations, we have designed a new large area HFM by utilizing plating and photoetching techniques and then developing a calibration method that is capable of calibrating several HFM simultaneously. This paper describes the design of the HFM, its performance, and the method of calibration.

Structure and Performance

The detectors currently used in HFMs are generally thermopiles. The detector of our new HFMs is also a thermopile. Since most conventional thermopiles are prepared by winding metal wire, deposition, etc., its cost of construction is very high, and the preparation of large area HFMs is almost impossible. This is the main reason why we employed the techniques of both plating and the photoetching.

In applying these techniques, three problems arose; first, the combination of thermocouple materials was limited to pure metals such as plating materials. The combination of copper and nickel was selected. Although this combination has a relatively smaller electro magnetic force (emf) compared to other combinations such as copper and constantan, this disadvantage was overcome by using a large number of thermocouples arranged in a large area. The second problem was how to connect the cold and hot junctions in a series at opposite surfaces.

Fortunately the inner wall of the "through holes" formed in a substrate could be plated. Therefore, it was possible to connect electrically both surfaces of the substrate.

Last, it was necessary to construct the large area HFMs and to cut whole HFMs into the arbitrary sizes depending on particular measuring applications. The construction of the large area HFM is so designed that it can be used not only as a large area HFM but also as a small area HFM without any modification by cutting the former into small units. It is, therefore, possible to put them on a small area of a wall (such as corners) by cutting the large area HFM into small ones when the original one is too large to be installed.

As shown in Fig. 1*a*, a pattern design containing a canal solved this problem. In the example presented in Fig. 1*a*, the unit's pattern is 75 by 75 mm^2,

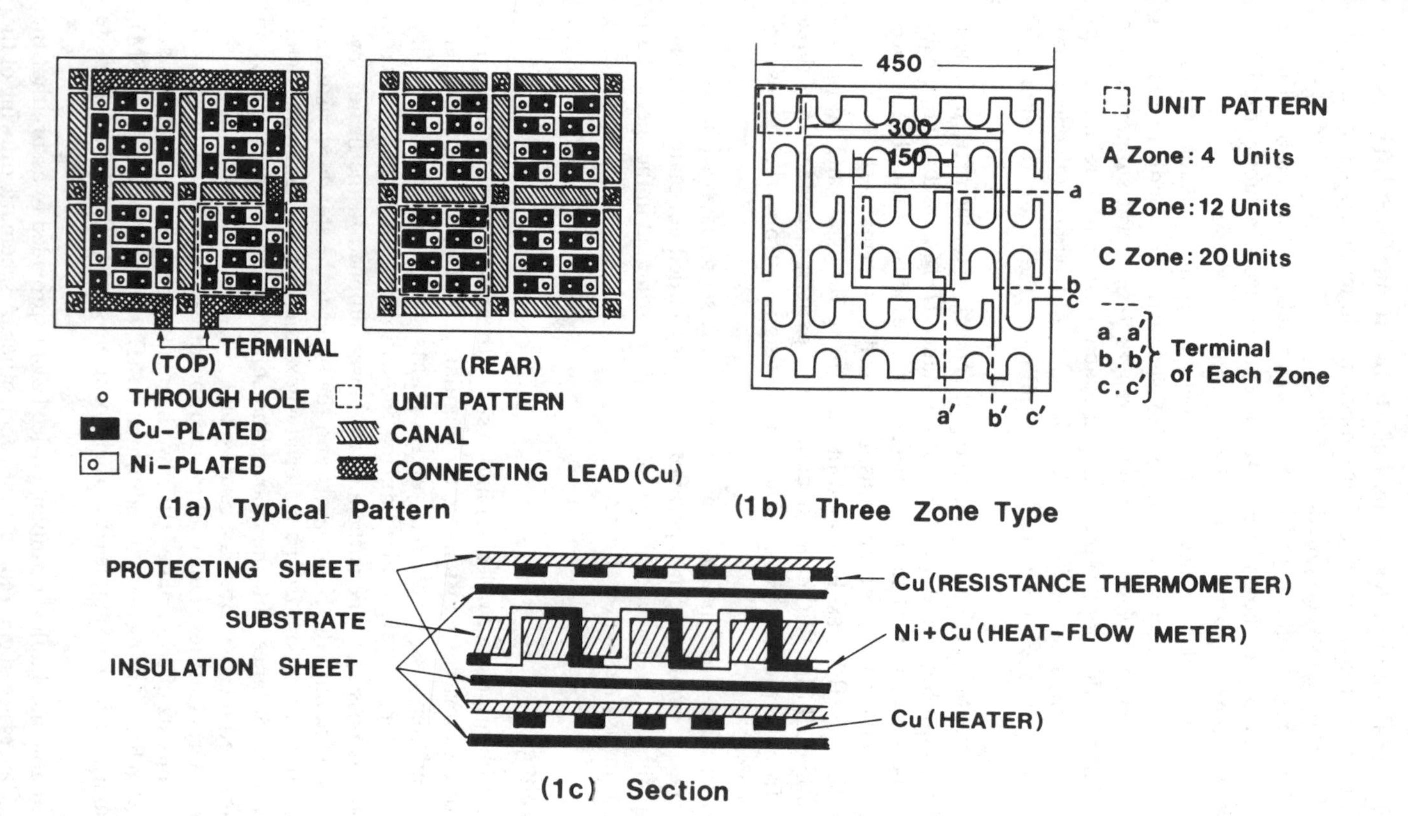

FIG. 1—*Structure of heat flow meter.*

and 16 (4 by 4) a unit pattern forms one 310 by 310 mm^2 HFM. The thermopile in each unit consists of 288 thermocouple junctions. As a result, the whole HFM contains 4608 junctions.

The temperature sensor and heater layer are also developed by utilizing similar techniques as shown in Fig. 1*c*. The three layers including the HFM can be also combined with each other by suitable processing.

An electrical insulation sheet is placed between them. The substrate which serves as a thermal barrier should be also electrically nonconductive, and it also must possess sufficient mechanical strength to serve as a support for the large number of thermopile junctions. For example, glass-fiber reinforced epoxy resin and the like can be used to form a plate having a thickness from about 0.4 mm to several millimetres depending on the required response of the HFM.

The number of junctions can be changed depending on the pattern design; therefore, both high- and low-thermal resistance HFMs are available, each having the same order of response because the response of the HFM is proportional to the number of junctions multiplied by thermal resistance.

A zone-type HFM, as illustrated in Fig. 1*b*, serves as a standard to calibrate other HFMs. Within its structure, the pattern is divided into three zones from center to edge. The output signals from the zones are independent.

Furthermore, when the heater and temperature sensor layer with the same zone type are combined, the operation is like a guarded hot plate. From another point of view, the lateral heat losses can be estimated in the special case of one dimensional heat transfer.

Calibration

The calibration was carried out in two steps: the first step was the establishment of a calibration constant for the standard HFM with three zones; and the second step was to calibrate several HFMs simultaneously. In the first step, two references are utilized during the calibration (it means that cross checking is possible): one is the electric power supplied to the heater layer that is bonded onto the HFM layer, and the other is the standard insulation sample that was cut out from a material whose thermal resistance was determined in accordance with the guarded hot plate method. The former is an absolute method; the latter is a relative one.

The general features of the calibration apparatus are illustrated in Fig. 2. It mainly consists of a hot plate (HP), two heater layers (PH), insulation material (IS), a heat flow meter to be calibrated (HM), a thermocouple sheet (SM), a standard insulation sample (RS), thermocouple sheet (SB), and a cold plate (CP).

A suitable mechanical clamping device (CL) is provided to assure that the various layers are in good contact with each other. A schematic drawing of the

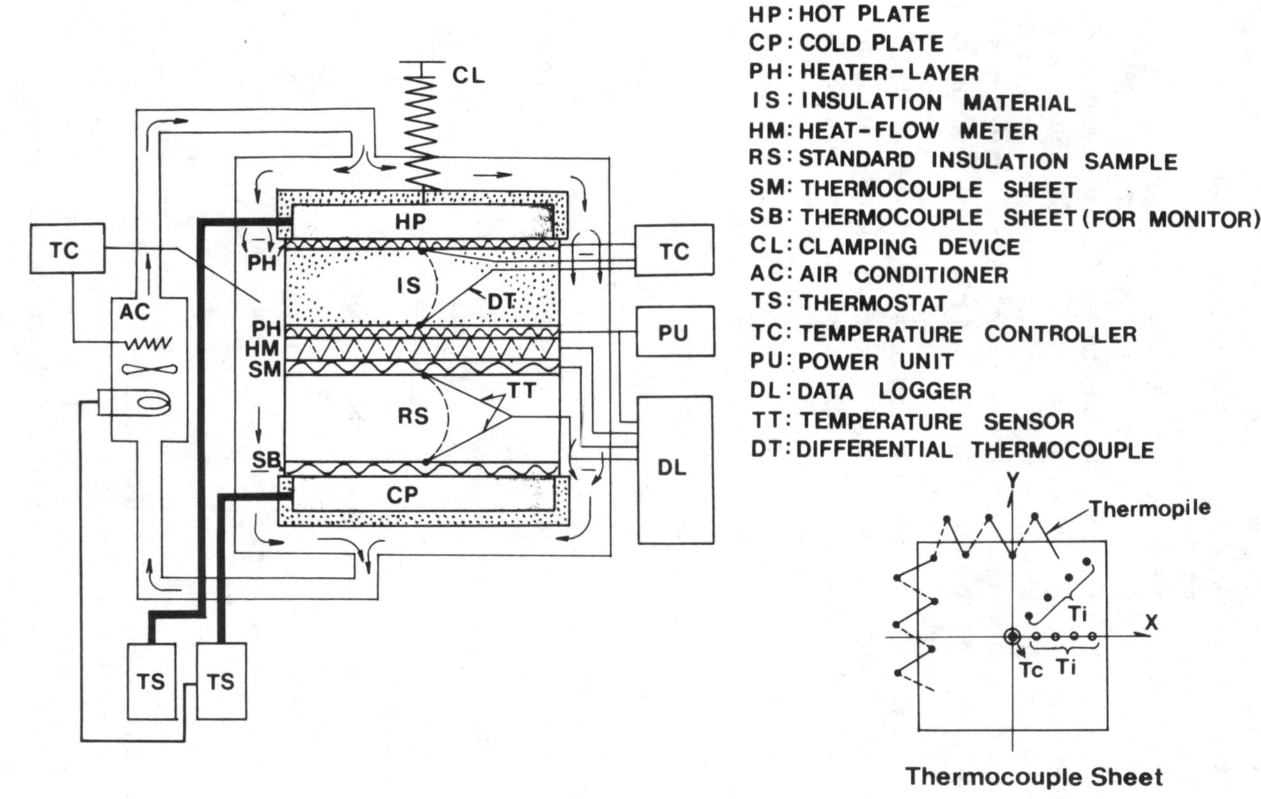

FIG. 2—*Schematic drawing of the calibration apparatus.*

SM is shown at the bottom of Fig. 2. It acts as the most important device on which 10 thermocouples are mounted along a diagonal and x-axis directions to measure the temperature distribution on the surface of the HFM during the calibration. In addition, 12 embedded differential thermocouples in the SM as per illustration in Fig. 2 are connected in a series to form a thermopile between the ambient air and the edge.

One set of junctions for this thermopile is located 10 cm distance away from the edge of the surface, and the other set of junctions is located in the air curtain. Namely, the temperature difference between them is detected, and it is used to control the edge heat loss in the second step of the calibration. An air conditioner (AC) is provided not only to control but also to maintain the ambient air at an arbitrarily constant temperature required for the first step of the calibration. Another SB is placed on the CP to monitor the same phenomenon supplementally.

Two junctions of a DT are placed on both surfaces of an IS as per Fig. 2 shows. If temperatures of both IS surfaces are controlled at the same temperature so that an output from the DT may be zero, the power supplied to PH reaches CP only, through HM and an RS.

In the first step of the calibration, the surrounding air temperature is fixed at higher or lower temperatures compared to the temperature Tc at the center of the SM.

At that time, a group of temperatures Ti at the specified positions along a diagonal and x-axis direction are measured (see Fig. 3). As the result, the summed temperature difference $\Sigma Ti - Tc$ is obtained. Furthermore, if this quantity is near zero, the edge heat loss is approximately zero. Since a one-dimensioned heat flow could not be exactly experimentally attained, the edge heat loss had to be determined graphically.

As an example, Fig. 4 gives the calibration constant corresponding to various $\Sigma Ti - Tc$. The black circular marks represent the calibration constants based on the standard insulation sample (relative method) and the white ones represent those based on the power measurement (absolute method). According to our test results showed in Fig. 4, the cross point of the two curves is very close to the line of $\Sigma Ti - Tc$ which shows zero (line of $\Sigma Ti - Tc = 0$), and the deviation is approximately less than 1%. In other words, the calibration constants obtained from the two methods are almost the same, so that we can see that the constants are correct.

In the second step of the calibration, a group of a maximum of ten HFMs replaces the RS in the first step, and the surrounding temperature is controlled so that the output of the thermopile between ambient air and the edge is equal to zero. Under this condition, the output signals for each HFM to be calibrated, the power supplied to the PH, or the output of the HFM calibrated at the first step are recorded, and then the calibration constants are determined.

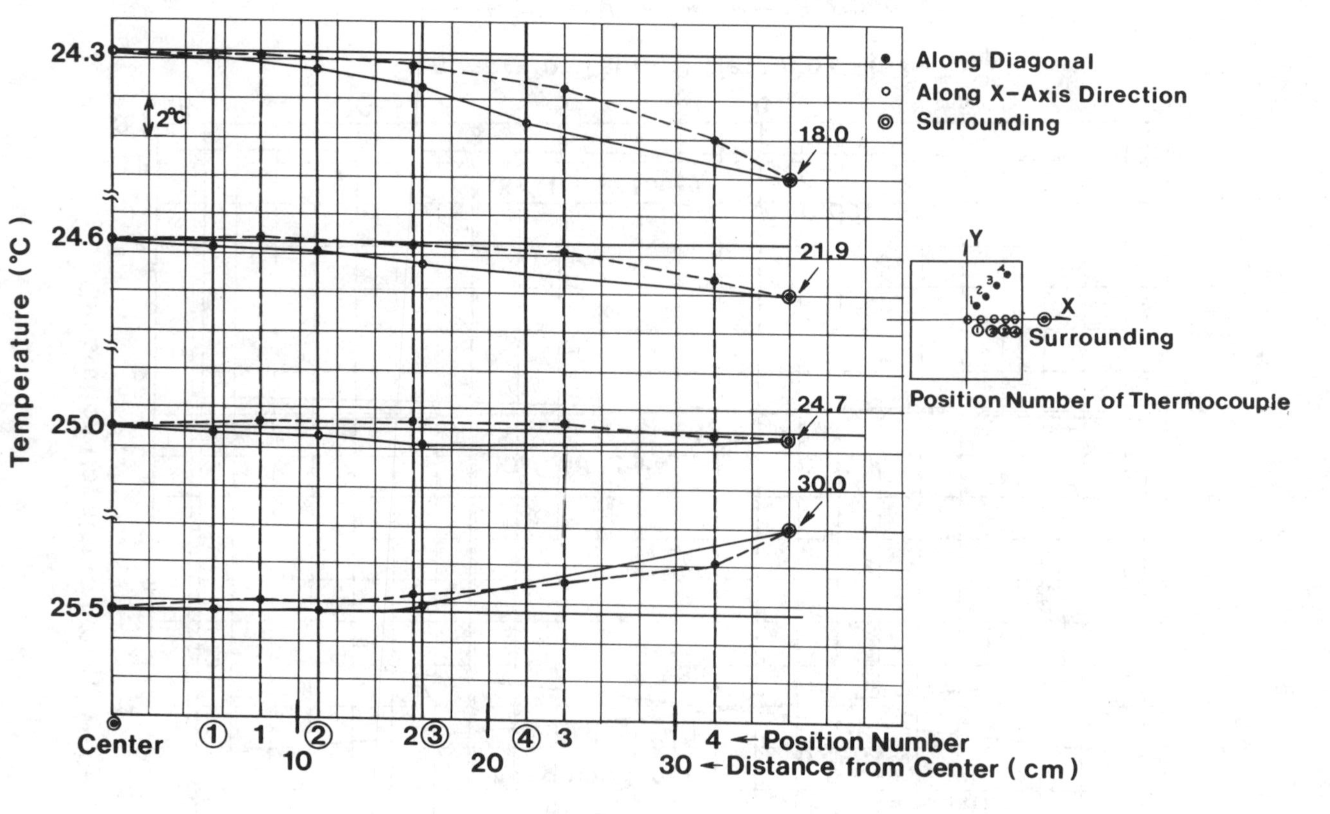

FIG. 3—*Surface temperature distribution.*

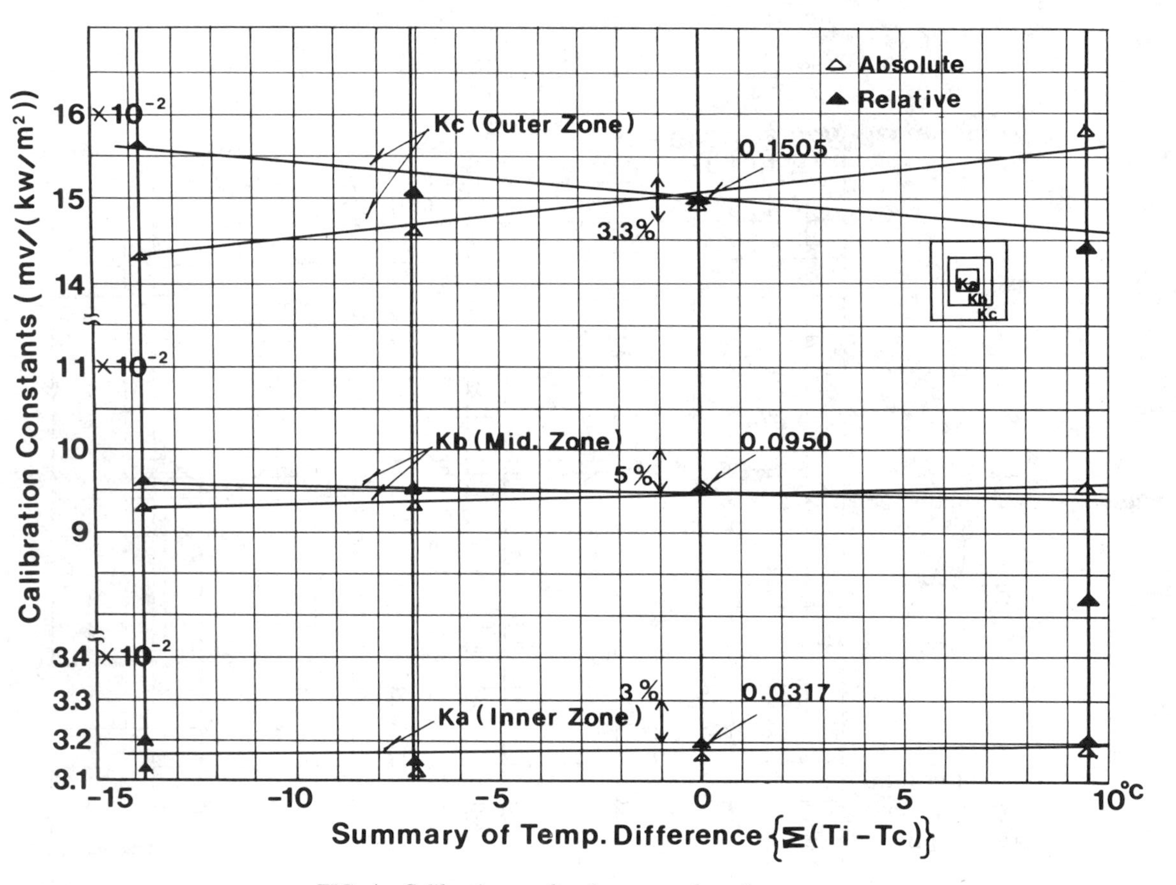

FIG. 4—*Calibration results of zone-type heat flow meter.*

Conclusion

A newly developed large area HFM having a pattern that can be divided into smaller ones was fabricated by utilizing photoetching and plating techniques. The inner wall of the "through hole" penetrating through the substrate was plated to connect electrically both surfaces of the substrate.

Calibration is accomplished by two independent methods which are based on the known thermal resistance of a standard insulating sample and the power measurement. Thermocouple sheets are introduced in order to find the most ideal condition and also control the surrounding temperature. The experimental uncertainty is about $\pm 1\%$.

There still remains a few problems to be solved. For example, the determination of temperature coefficient, etc.

Acknowledgment

The authors are indebted greatly to T. Kobayashi and Y. Fujii for their fabrication techniques. It also is a pleasure to acknowledge the experimental work of Mr. Kawabata and T. Aoshima.

Reference

[*1*] Klems, J. H. and Dibartolomeo, D., *Review of Scientific Instruments*, Vol. 53, No. 10, Oct. 1982, p. 1611.

Nathaniel E. Hager, Jr.[1]

Recent Developments with the Thin-Heater Thermal Conductivity Apparatus

REFERENCE: Hager, N. E., Jr., **"Recent Developments with the Thin-Heater Thermal Conductivity Apparatus,"** *Guarded Hot Plate and Heat Flow Meter Methodology, ASTM STP 879*, C. J. Shirtliffe and R. P. Tye, Eds., American Society for Testing and Materials, Philadelphia, 1985, pp. 180–190.

ABSTRACT: A thin-foil heating element is employed in a thermal conductivity apparatus that operates in accordance with the basic concept of the guarded hot plate. The metal foil is sufficiently thin that lateral heat flow along the plane of the heater is insignificant, at least in the central region, so there is no need for isolation and separate temperature control of a guard region. This makes the apparatus simple and inexpensive to construct and operate. The resultant low-mass configuration minimizes drift error and usually reaches steady state in less than 30 min with 0.5-cm-thick specimens of nonmetallic solids. Thermal conductivity values are calculated on an absolute basis with an estimated accuracy of $\pm 3\%$.

Speed and simplicity make the apparatus well suited for product-development and quality-control applications. Because temperature differences within the sample do not exceed a few degrees, uncertainty in mean sample temperature is sharply limited, water-vapor-migration problems are minimized, and thermal conductivity can be measured close to melting points or other critical temperatures. New data obtained with dry calcium silicate give preliminary experimental verification of the validity of the apparatus in the temperature range extending between ambient and 550 K.

KEY WORDS: thermal conductivity, equipment, measurement, nonmetallic solids, calcium silicate

The guarded hot plate (GHP) apparatus is usually regarded as the principal device for thermal conductivity measurement [*1*]. As shown earlier, application of the GHP principle can be greatly simplified by using a thin metal-foil heater as the hot plate [*2–5*]. With this configuration, edge losses can be kept small without guarding, so that construction and operation of the appa-

[1]Senior research associate, Research and Development, Armstrong World Industries, Inc., Lancaster, PA 17604.

ratus are less complicated. Although not capable of the highest degree of accuracy attained with the GHP, the thin-heater apparatus is estimated to be capable of 2% accuracy.

Details of the thin-heater method and apparatus are given in the previously cited papers. The purpose here is to provide a review and describe a specific apparatus developed at this laboratory for measuring thermal conductivity of nonmetallic solids at mean temperatures ranging from ambient to 550 K.

Concept

The thin heater consists of a uniform strip of metal foil folded in half with an intervening layer of electrical insulation, as shown in Fig. 1. The ends of the foil are established as equipotentials by attaching heavy metal buss bars. The folded foil configuration is sandwiched between two identical specimens of sample material. The specimen thickness does not exceed one tenth the least dimension of the heater. Two heavy heat-sink plates enclose the whole assembly.

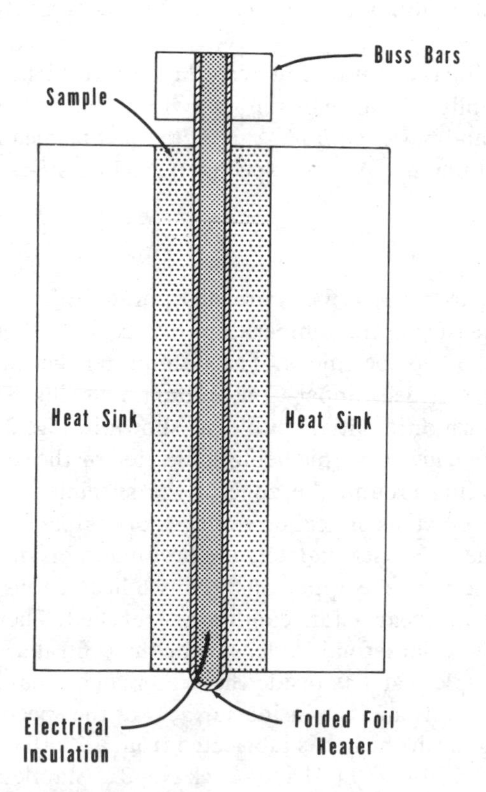

FIG. 1—*Schematic view of basic configuration of thin-heater thermal conductivity apparatus, including specimens and heat sink blocks.*

A power supply is connected to the buss bars, and a constant current is passed through the foil, causing the latter to heat. Assuming the temperature of the heat-sink blocks stays fixed, a steady-state balance is eventually reached such that the rate of heat generation in the heater equals the total rate of heat flow through the two specimens.

Under this condition, at least near their central portions, the thermal conductivity of the specimens can be judged by observing how large a temperature difference is needed between heater and heat sink to propagate heat flow through the specimens at the generated rate. The thermal conductivity is given by the equation

$$k = FD/\Delta T \quad (1)$$

where

k = thermal conductivity, W/m K,
F = heat flow per unit area through specimen, W/m^2,
D = sample thickness, m, and
ΔT = temperature difference between heater and sink, K.

To make a practical apparatus in accordance with this concept, a procedure must be established for determining F from knowledge of the parameters of the heater and power supply. And there must be means for measuring the temperature difference ΔT and specimen thickness D.

Assembly

The specific apparatus described here incorporates differential thermocouple systems for measuring the temperature difference ΔT between opposite faces of each of the two specimens. The differential thermocouples are cut from a strip of edge-welded copper-Constantan foil as shown in Fig. 2*a*. Each specimen has its own differential couple, as shown in Fig. 2*b*, with platelets on opposite sides of a given sample being connected by the continuous strip of Constantan extending around the edge of the specimen, and with copper leads extending to provide connection to the external readout instrument. As explained next, the two inner platelets are laminated inside the heater. (Because of the symmetry of the apparatus, the two heater foils are at the same temperature when the steady-state condition is reached. Therefore, the space between these foils is isothermal, at least near the central area of the heater, and the thermocouple platelets inside the heater foils should be at the same temperature as the foils and the inside surfaces of the specimens.)

As shown in Fig. 3, the heater is fabricated from a 10.16 by 30.48-cm (4 by 12-in.) piece of 0.00254-cm (0.001-in.)-thick No. 321 stainless steel foil.[2] Two

[2]The apparatus described was designed using English units.

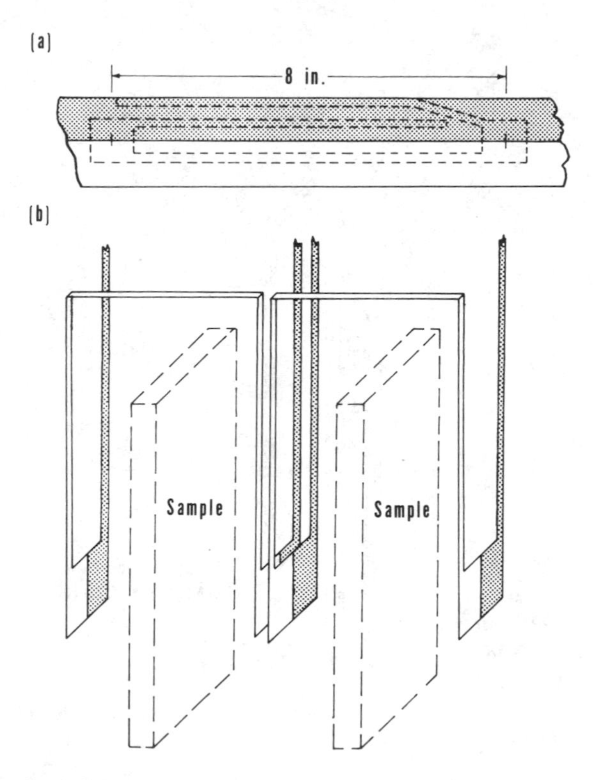

FIG. 2—*Details of differential thermocouple configuration with copper foil shaded and constantan foil unshaded; part* (a) *shows layout* (not drawn to scale) *for cutting differential couple from 2.54-cm (1-in.)-wide strip of edge-welded copper-constantan ribbon; part* (b) *shows relative position of platelets in apparatus.*

10.16 by 15.24-cm (4 by 6-in.) sheets of 0.0254-cm (0.010-in.)-thick silicone glass laminate are laminated together, enclosing two of the thermocouple platelets as described previously. These platelets and their leads are electrically insulated from each other with 0.00508-cm (0.002-in.) polyimide film. The centers of the platelets should be located about 6.35 cm (2.5 in.) from the fold in the heater, and midway between the heater edges.

The four leads from the two platelets extend from between the two silicone glass sheets near the middle of one of the 10.16-cm (4-in.) edges. All bonding is done with pressure-sensitive silicone adhesive. The heater is then folded over and bonded to the silicone glass assembly, using the same adhesive, so that the buss bars are at the same end of the silicone glass as the leads from the thermocouple platelets. A 2.54 by 10.16-cm (1 by 4-in.) mounting strip made of 0.645-cm (0.25-in.)-thick silicone glass laminate is fastened to the buss bars with binding posts, bolts, and nuts, with the thermocouple leads extending through a slot in the mounting strip. The two outer thermocouple platelets are positioned so that their centers are about 6.35 cm (2.5 in.) from

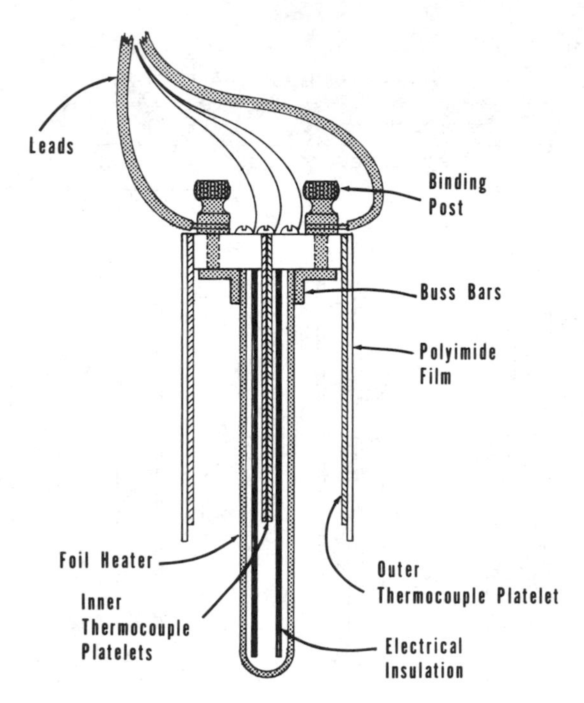

FIG. 3—*Sketch of thin-heater unit, not including specimens and heat sink blocks; horizontal scale expanded to show thin foils and insulation sheets.*

the folded end of the heater, and their leads are then bonded to the side of the mounting strip. Their centers are aligned with the centers of the pair of platelets inside the heater. The two copper leads extending from the between-heater platelets are folded together and clamped under a nylon screw tapped into the mounting strip. A thermocouple quality copper wire, also joined in this connection, extends to the readout instrument. The copper leads from the outer platelets are similarly connected to external wires using two other screws. Power leads are fastened to the binding posts. These are bound together with the thermocouple wires to form a cable approximately 91.44 cm (36 in.) long.

It is found helpful to reinforce the outer thermocouples with films of 0.00508-cm (0.002-in.) polyimide film. The films are cut large enough to cover platelets, leads, and a portion of the side of the mounting strip. These films are mounted with adhesive to the outer surface of the foils so they do not interfere with the thermal contact between the thermocouple platelets and the specimen. Note that the outer thermocouples, including the polyimide film covers, can be bent to accommodate variations in specimen thickness.

After the specimens are placed between the heater and the outer thermocouple platelets, the assembly is clamped between two 10.16 by 15.24-cm (4 by 6-in.) blocks of 2.54-cm (1-in.)-thick aluminum. The clamping is done with spring clamps, with C-clamps, or by an overall wrapping of tape. Note that the polyimide film on the outer platelets prevents them from being electrically shorted by the aluminum heat-sink blocks.

Operation

The typical procedure is to cut two 10.16 by 15.24-cm (4 by 6-in.) specimens of material having thickness in the range from 0.254 to 0.645 cm (0.10 to 0.25 in.). After thickness measurements are completed and the specimens are placed against the two sides of the heater with the outer thermocouples outside the specimens, the insulated heat sink blocks are positioned and the whole assembly is clamped. (If the samples are compressible, their actual thickness during measurement can be determined by measuring the distance between the heat sink blocks and subtracting out the heater thickness.)

The apparatus is then connected to the power supply and readout equipment as shown schematically in Fig. 4. A microvoltmeter, such as the Keithley Model 150-B, is used for the thermocouple readout. An all-copper switch is included to facilitate reading the temperature differences between faces of both specimens. If desired, the microvoltmeter output can be fed to a digital readout device. The power is delivered with a regulated d-c power supply capable of delivering currents of the order of 5 A. (A standard 2-V lead-acid storage battery has been used satisfactorily for powering the heater.) A precision d-c ammeter with about 0.5% accuracy is suitable for measuring the heater current.

To obtain a thermal conductivity value for the sample at a temperature

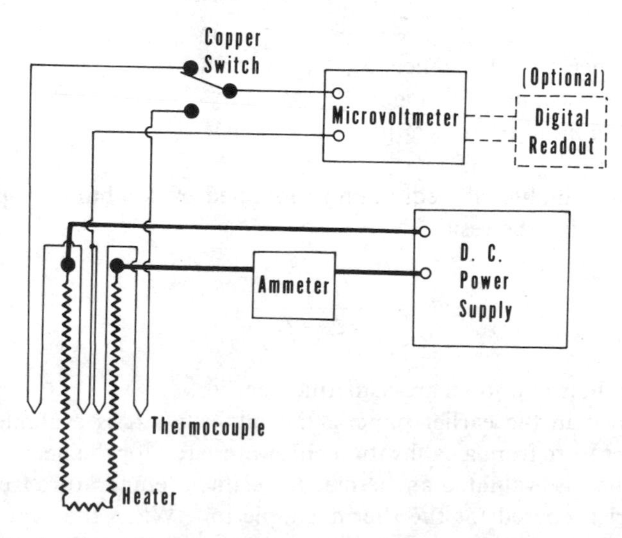

FIG. 4—*Electrical circuit diagram, including readout and power supply.*

near room temperature, the thin-heater assembly is simply stood upright on the laboratory bench. The power supply is turned on. The thermocouple readout is watched until readings at 5-min intervals remain unchanged. This usually takes 15 to 30 min when using 0.645-cm (0.25-in.) specimens of cork compositions. At this point, values are recorded for the heater current, specimen thicknesses, thermocouple electromotive force, and heat sink temperature.

Calculations

The specimen heat flux is calculated from the equation

$$F = (\rho/t)(i/w)^2 \tag{2}$$

where

F = heat flux, W/m^2,
ρ = foil electrical volume resistivity, ohm cm,
t = foil thickness, cm,
i = heater current, A, and
w = heater width, m.

The temperature difference between specimen faces is given by the equation

$$\Delta T = c\Delta e \tag{3}$$

where

ΔT = temperature difference, K,
c = thermocouple coefficient, K/μV, and
Δe = differential thermocouple readings, μV.

The thermal conductivity equation is obtained by combining Eqs 2 and 3 with Eq 1, yielding the result

$$k = \frac{(\rho/t)(i/w)^2}{c\Delta e/D} \tag{4}$$

where all symbols and units are as defined earlier.

As described in the earlier papers, the quantity ρ/t is evaluated at the heater temperature from a calibration chart prepared for the heater foil, and the coefficient c is evaluated at the mean specimen temperature from a calibration chart prepared for the thermocouple foil. When the ratio $\Delta e/D$ is different for the two specimens, the average is inserted into Eq 4.

It is to be noted that all quantities in Eq 4 are determined by direct physical measurement. There are no arbitrary factors requiring calibration by comparison with a standard material—that is, the apparatus yields results on an absolute basis. It can be viewed as a form of GHP apparatus, with the hotplate too thin to conduct heat laterally from the central zone with the result that, within that central zone, all heat flows outward through the specimens in the ideal manner assumed. Because of the thinness of the foil, there is no need for a gap between the central and peripheral heater regions acting, respectively, as hot plate and guard.

Characteristics

Ease and Low Cost of Fabrication

Experience here shows that one man can construct one or two thin-heater devices per day, thus making it practical to have a substantial number of such units on hand; this leads to the following advantages:

1. A number can be run simultaneously, using a single thermocouple readout system, thus boosting operator productivity.
2. Downtime is eliminated in the event of damage or destruction of a unit due to chemical attack or other effect from the sample material.
3. Intramural round robins can be run with standard specimens to maintain confidence that one or more units have not developed problems. This capability is illustrated in Fig. 5, where results are shown for a pair of specimens measured in four different thin-heater units.

Specimen Requirements

The apparatus described is suitable for measuring homogeneous nonmetallic solids and many types of thermal insulating materials. Specimen thickness should not exceed one-tenth the least dimension of the heater; with the 10.16 by 15.24-cm (4 by 6-in.) unit described, the specimen thickness should not exceed 1.016 cm (0.4 in.). With coarsely textured or infrared transparent materials, enclosing surfaces should have a high-emissivity coating, and consideration should be given to using a larger thin heater in combination with larger specimen thickness.

Temperature Range

The described heater specimen assembly can be operated on the lab bench without special control of room temperature or use of a temperature-controlled enclosure. The mean specimen temperature is determined by measuring the temperature of the heat sink blocks and adding $\Delta T/2$. Alternately,

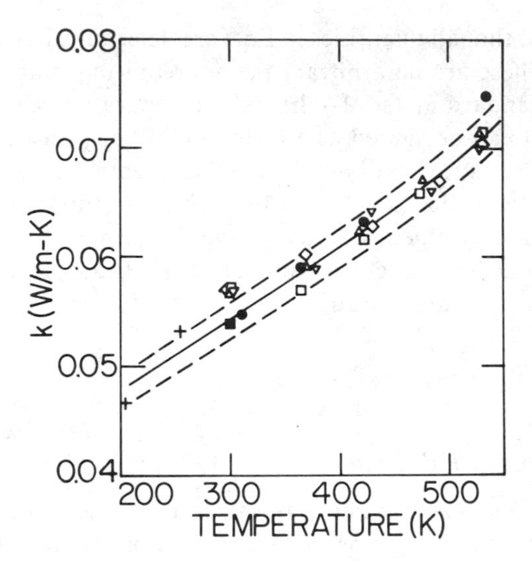

FIG. 5—*Experimental thermal conductivity data for calcium silicate; blackened circles represent manufacturer's data published for 0.21 g/cm³ calcium silicate product; triangles and squares represent four series of measurements performed on same product, each series being done on different thin-heater device; blackened square represents repeat of room-temperature measurement obtained after specimens dried during higher temperature measurements; plus signs represent Speil's low temperature data for calcium silicate at same density; dashed-line curves are drawn 3% above and below solid-line curve estimated to best fit thin-heater data for dry calcium silicate.*

the assembly is enclosed in a Teflon or Mylar bag and immersed in an ice or cryogenic bath to obtain low temperatures. Current work is being done in ovens as high as 600 K. Results obtained for calcium silicate between ambient temperature and about 550 K are shown in Fig. 5. Although some materials used in the described thin-heater device deteriorate rapidly at higher temperatures, recent work with synthetic mica electrical insulation and platinum-platinum rhodium thermocouple ribbon has yielded measurements up to about 800 K.

Small ΔT

Most work here is with ΔT-values ranging between 2 and 5 K. Using a commercial nanovoltmeter to read the differential thermocouple signal, it is feasible to measure thermal conductivity with ΔT as low as a few hundredths of a degree. Under such conditions, determination of the mean specimen temperature can be made with little error, even when the temperature varies nonlinearly between the two specimen faces. Use of small ΔT minimizes water vapor migration in moisture-containing specimens. It also makes practical the measurement of thermal conductivity near melting points or other phase changes.

Speed

Steady state is usually reached within about 10 to 30 min. with specimens in the 0.254 to 0.762-cm (0.10 to 0.30-in.) thickness range. Because settling is so fast, the heater so light, and the heat sink so massive, normal room-temperature or oven-temperature variations have not been observed to have a significant effect on measurements.

Precision

Extended series of measurements with a given thin-heater unit, all performed with the same set of specimens, have yielded results consistent to within about 2%. Data obtained for one pair of specimens in four different thin heaters, shown in Fig. 5, typify the reproducibility usually observed. In each series, the individual points fall no more than 2% above or below the best-fit curve. (An apparent exception indicated for the room-temperature data is explained by the fact that moisture content affected the result at low temperatures; when the specimen set was dried during high-temperature measurements, the discrepancy was removed.)

Accuracy

In view of the thickness-precision of metal foil obtained thus far, it is estimated that the highest absolute accuracy to bc expected is of the order of 2%. As shown in the cited earlier papers, this estimate is not inconsistent with results of experimental tests conducted to compare the thin-heater device with conventional apparatus.

Several units of the present thin-heater device were studied by comparing thermal conductivities obtained for two standard corkboard materials with values obtained for the same materials in accordance with the ASTM Test for Steady-State Thermal Transmission Properties by Means of the Heat Flow Meter (C 518). With all measurements made at the same mean specimen temperature, approximately 297 K, average thin-heater results fell consistently within 3% of results obtained with ASTM Method C 518.

In Fig. 5, thin-heater data for one brand of calcium silicate insulation (density = 0.21 g/cm^3) are compared with data published by that same material's manufacturer in compliance with ASTM Standard Specification for Calcium Silicate Block and Pipe Insulation (C 533). Although agreement between thin heater and published data is excellent between 300 and 450 K, a difference approaching 5% is evident at the extreme upper end of the temperature range included. Figure 5 also shows that low-temperature GHP data obtained by Speil [*6*] for calcium silicate of the same density are consistent with downward extrapolation from the thin-heater data.

Simplicity

Complete elimination of the guard-control system greatly simplifies both fabrication and operation of the apparatus. It also eliminates conceptual difficulties associated with the gap between guard and hot plate. The light weight of the heater minimizes impact of the "drift error" sometimes of concern when using heavier hot plates.

Applications

The low-cost and high-productivity characteristics make the thin-heater apparatus well suited for quality control and product optimization work.

It is ideally suited to measurement of thermal conductivity of felts, wood products, ceramics, and polymers. It should be considered for measuring geological materials that ordinarily encounter moisture problems, and for obtaining data near transition temperatures. It is also used for calibration of heat-flux sensors.

Acknowledgment

The author thanks F. G. Dochat for his careful work in constructing and operating thin-heater devices of the type described here.

References

[1] *1983 Annual Book of ASTM Standards*, American Society for Testing Materials, Philadelphia, 1983, Vol. 04.06, pp. 21-54.
[2] Niven, C. and Geddes, A. E. M., *Proceedings*, Royal Society (London), Vol. A87, 1912, p. 535.
[3] Hager, N. E., Jr., *Review of Scientific Instruments*, Vol. 31, No. 2, Feb. 1960, pp. 177-185.
[4] Hager, N. E., Jr., U.S. Patent No. 3,045,473, 24 July 1962.
[5] Hager, N. E., Jr., *Transactions*, Instrument Society of America, Vol. 8, No. 2, 1969, pp. 104-109.
[6] Speil, S. in *Thermophysical Properties of Thermal Insulating Materials*, Technical Documentary Report No. ML-TDR-64-5, Air Force Materials Laboratory, Wright Patterson Air Force Base, Ohio, April 1964, p. 62.

Methodology

Charles M. Pelanne[1]

Development of a Company Wide Heat Flow Meter Calibration Program Based on the National Bureau of Standards Certified Transfer Specimens

REFERENCE: Pelanne, C. M., **"Development of a Company Wide Heat Flow Meter Calibration Program Based on the National Bureau of Standards Certified Transfer Specimens,"** *Guarded Hot Plate and Heat Flow Meter Methodology, ASTM STP 879*, C. J. Shirtliffe and R. P. Tye, Eds., American Society for Testing and Materials, Philadelphia, 1985, pp. 193-205.

ABSTRACT: The development of a company-wide heat flow meter calibration program based on the National Bureau of Standards Certified Transfer Specimens is discussed. A description of the procedure followed in order to ensure accurate measurements of thermal insulation in conformance with the Federal Trade Commission's requirements is given. The procedure ensures the maintenance of the equipment calibration for all testing locations by providing each with a set of secondary reference specimens. To stress the importance of the program, the reference specimens and the maintenance of the equipment calibration, a Certificate of Calibration for each specimen and a set of instructions was provided to each test location.

KEY WORDS: thermal insulation, thermal resistance specimens, certified transfer specimens, heat flow meter apparatus, calibration, standardization, quality assurance, thermal conductivity measurements, Federal Trade Commission, National Bureau of Standards, glass fiber insulation, mineral fiber insulation

In April 1981, the National Bureau of Standards (NBS), Gaithersburg, Maryland, provided certified transfer specimens (CTS) to industrial and commercial laboratories. These specimens made of low-density fibrous insulation batts were necessary to ensure uniform implementation of the full thickness testing requirements of the Federal Trade Commissions' (FTC)

[1]Research associate-retired, Manville R/D Center, Denver, CO 80217; present address: 4900 Pinyon Drive, Littleton, CO 80123.

trade regulation rule [*1*] on the labeling and advertising of home insulation. Measured by means of a new NBS line heat source guarded hot plate (LHSGHP), the specimens were supplied initially to the laboratories having submitted purchase requests. A photograph of the NBS 1000 mm LHSGHP is shown in Fig. 1. The CTS would provide each testing facility with a single reference to an official source.

The implementation of full thickness testing required a number of important steps subsequent to the reception of the CTS. The initial steps undertaken by the Heat Transfer Laboratory of the Manville Research and Development Center are outlined.

Objectives

The objectives of this program were to:

1. Calibrate the Research and Development Center (R&D) and plant Quality Assurance department (QA) thermal resistance measurement equipment against the CTS provided by NBS.
2. Provide secondary reference specimen for the calibration and maintenance of calibration for the R&D and QA equipment.
3. Provide data required to ensure fulfillment of the FTC requirements.

FIG. 1—*Photograph of the NBS 1000 mm line source GHP.*

Background

Prior to the availability of NBS CTSs, the Manville Corporation thermal measurement calibration system was based on the NBS standard reference material (SRM 1450) (96 to 128 kg/m^3) [*2*] and an internally developed calibration system referenced to the NBS guarded hot plate (GHP). This system, initiated in 1974, consisted of lower density specimens (11.2 to 33.6 kg/m^3) made from glass fiber insulation batt material [*3*]. Both of these materials were tested at 25.4 mm in the NBS "8 in." GHP. They were insufficient for the fulfillment of the testing requirements of the new FTC rule [*1*].

The CTS supplied by NBS were made from selected pieces of specially manufactured glass fiber insulation batt material at a density of about 9.6 kg/m^3 (0.6 lb/ft) of two thicknesses 25.4 mm (1 in.) and 76.2 mm (3 in.). This material was manufactured by the Manville Building Materials Corporation under very careful conditions in an attempt to achieve uniformity among the specimens. The required level of uniformity was achieved by careful visual selection of the individual pieces. In view of the implementation of this program, the Manville Research and Development ordered a quantity of the same material produced during this special run. Thus, all of the pieces of insulation prepared as secondary reference specimens, identified as Thermal Resistance Reference Specimen (TRRS), would be made from essentially the identical material as the primary CTS.

Prior to the initiation of this program, an experimental and mathematical study of the effect of thickness in this low-density glass fiber insulation was undertaken. The results of this study, which showed an excellent agreement between (1) our theoretical expectations, (2) the measurements performed on our stock material, and (3) the measurements on the CTS, were reported by Albers [*4*] at the 17th International Thermal Conductivity Conference in Gaithersburg, Maryland, in June 1981.

Calibration Program

NBS Certified Transfer Specimens (CTS)

In December 1980, Manville Service Corporation ordered from NBS two sets of specimens which were delivered early in May 1981. Each set consisted of one 25.4-mm (1-in.), one 76.2-mm (3-in.), and one 152.4-mm (6-in.)-thick specimens. The 152.4-mm specimen is made up of two 76.2-mm-thick pieces. Each of the six CTS had been tested individually in the NBS LHSGHP with upward heat flow. The test data were given on the certificate as measured for the test area of the NBS apparatus (406.4 mm in diameter) and for the test area of the apparatus in which they were to be used for calibration. In this instance, the data were given for the 254 mm square test area of the heat flow meter (HFM) (Modified "R-Matic"[2]) in which they were to be used. A detailed description of the process used by NBS to normalize the data is given in an NBS report by Rennex [5]. The normalized values were used for the calibration process. A photograph of the modified "R-Matic" HFM apparatus is shown in Fig. 2.

Reference Specimen Requirements

Our requirements were to provide each testing location with a set of specimens and several extra reserve sets of TRRS referenced to the NBS CTS. Two of these sets, consisting of matched pairs would be used for testing as pairs in a GHP.

To fulfill the requirements, fourteen sets of 609.6 by 609.6 mm (24 by 24 in.) specimens were prepared. Each set consists of three pieces of glass fiber insulation blanket, one 25.4 mm (1 in.) and two 76.2 mm (3 in.) thick. The specimens were evaluated individually, and the two 76.2-mm specimens combined to form a single 152.4-mm specimen.

Thus a total of four reference values were provided with each set of specimens: (*a*) one for the 25.4-mm piece, (*b*) two (one each) for the two 76.2-mm pieces, and (*c*) one for the combined 76.2-mm pieces tested at a 152.4 mm test thickness.

These specimens were identified as TRRS to differentiate them from the CTS of NBS.

Selection Process and Identification

The specimens were selected from 1219 by 2438 mm (48 by 96 in.) batts by visual observation over a large light box. The specimens were chosen from areas of the batt exhibiting the most uniformity of density. Particular attention was placed on the central test area.

Each specimen was weighed and identified with a code number placed on the lower right-hand corner of one face. This location identification serves as a guide for orientation of the specimen during test.

The density of each specimen was determined on the basis of the total specimen at the nominal test thickness; thus, it is known only as the bulk density. There is only need to refer to the specimen as a specific piece of material having a given thermal resistance. The determination of the test area density would require that the specimen center be cut out, thus making the specimen test area more vulnerable to damage. While of academic interest, maintaining the integrity of the specimen was considered more important. The weight of the specimen was determined and made part of the record in order to verify

[2]"R-Matic" is the trade name of a heat flow meter apparatus manufactured by the Dynatech R/D Co. of Cambridge, Massachusetts.

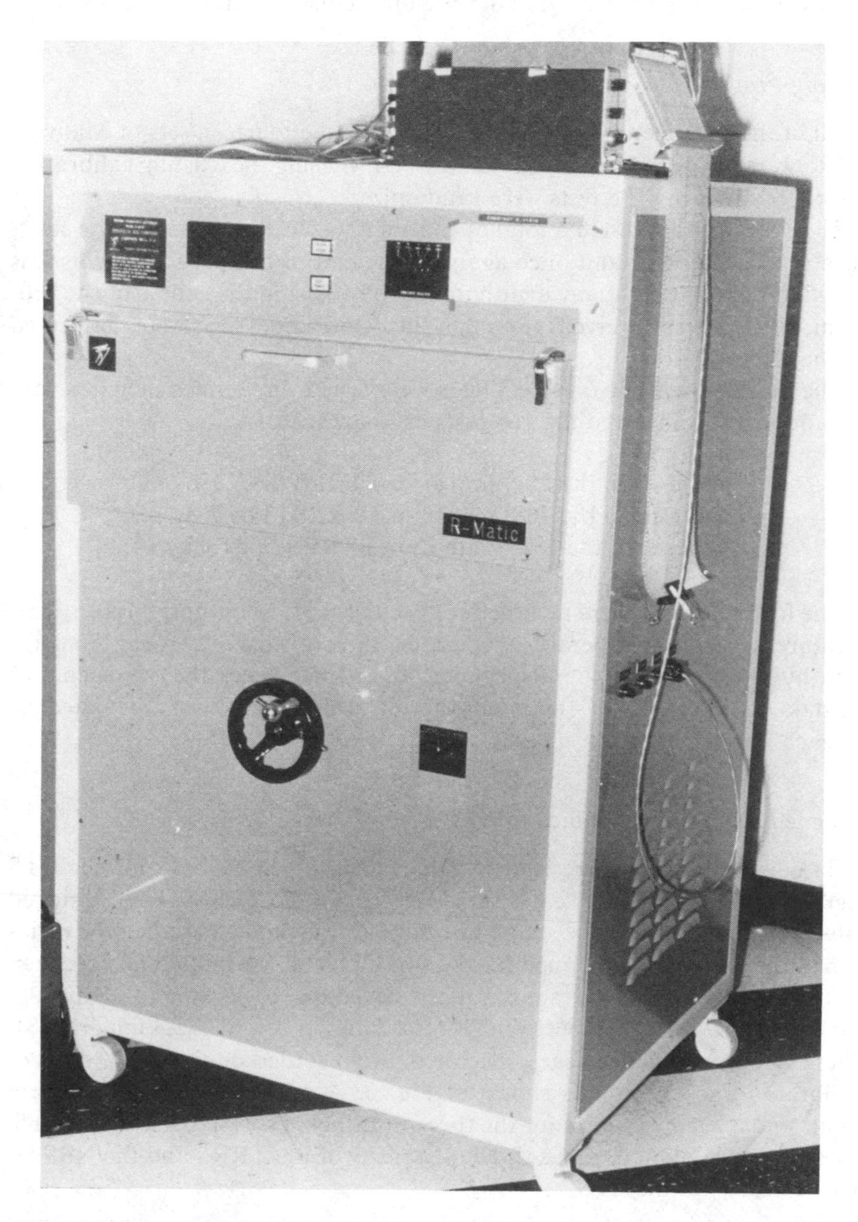

FIG. 2—*Photograph of the modified R-matic HFM apparatus used as the principal quality assurance thermal resistance determination apparatus in all Manville R&D and production locations.*

if the specimen remained physically in good condition. The data for the TRRS as shown in Figs. 3, 4, and 5 are on the basis of this bulk density.

Testing Program

All of the specimens, the six NBS CTSs and the fourteen sets of Manville TRRS were measured in a single "24 in." HFM using the existing calibration constant. All of the 198 tests were randomized.

The six CTS were tested a total of five times each, three times at the start, once at the midpoint and once again at the end of the program. This was aimed at determining if any drift had occurred during the course of the program. No drift was observed. A total of thirty individual tests were performed on the CTS.

The fourteen sets of Manville TRRS were tested three times each in a random order. This involved the 168 tests as follows:

42 tests on the 25.4 mm (1 in.) TRRS (3 by 1 by 14)
84 tests on the 76.2 mm (3 in.) TRRS (3 by 2 by 14)
42 tests on the 152.4 mm (6 in.) TRRS (3 by 1 by 14)

The following data were recorded for each test: HFM output, hot face temperature, cold face temperature, operator, time in-time out, room temperature, humidity, and weight of specimen. In order to check the consistency of the data, an interim "kα" (apparent thermal conductivity) was calculated after each test.

Assigned Thermal Resistance Values

Based on the thermal resistance values determined by NBS for the CTS thermal resistance values were assigned to each of the TRRSs. Each assigned value was provided with the standard error of the estimate based on the variation in the test results obtained for the two CTS and the individual specimen of a given thickness. (Average standard deviation, σ, in SI units, 0.00008, 0.00010, and 0.00025, respectively, for the 25.4, 76.2, and 152.4 mm TRRS). These values, assigned to each specimen, are recorded on the Certificates of Calibration, an example is shown in Fig. 6. A plot of the individual data points and a curve is shown for the three test thicknesses in Figs. 3, 4, and 5. These data are plotted versus the bulk density of the TRRS and the NBS reported density of the CTS.

Reference Specimens and Certificates of Calibration

Each test location was provided with one set of specimens consisting of three pieces of glass fiber insulation individually packed in special protective

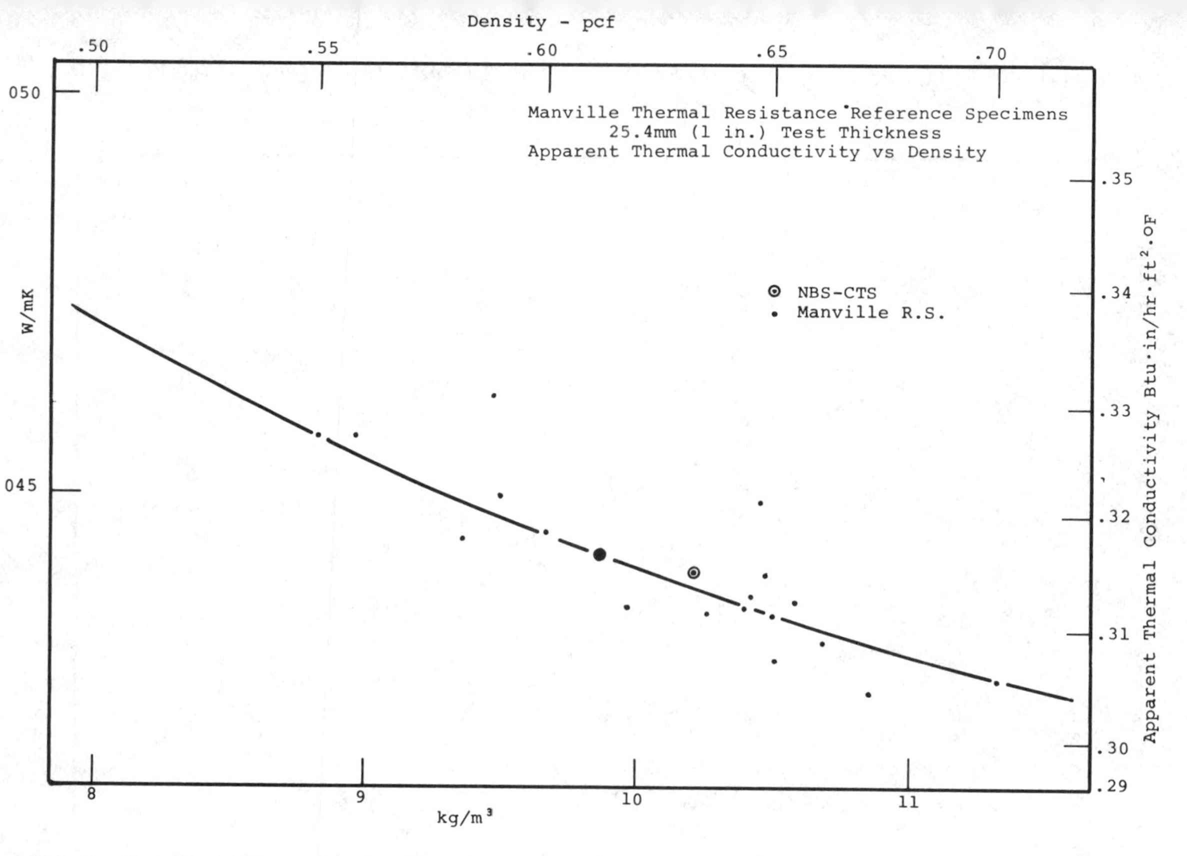

FIG. 3—*Plot of the CTS and Manville TRRS data points and regression curve for the 25.4 mm (1 in.) glass fiber reference specimens.*

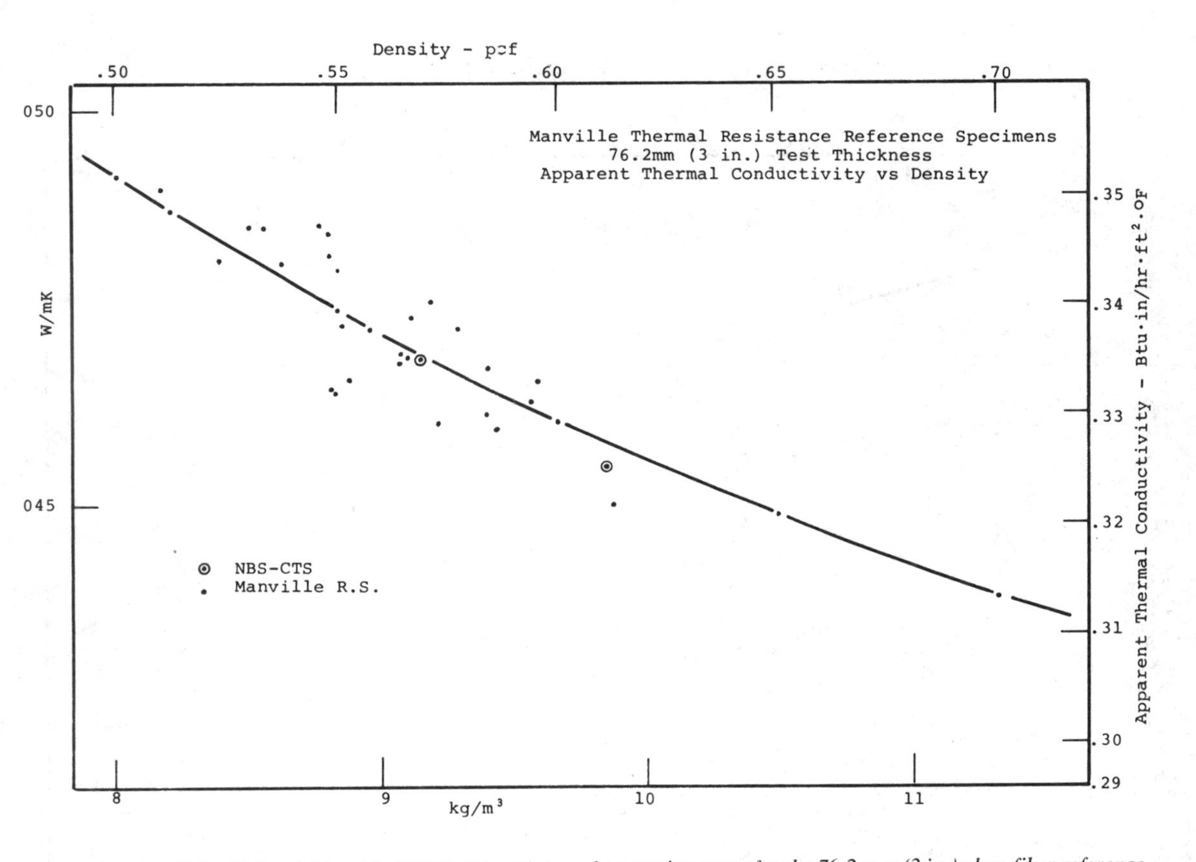

FIG. 4—*Plot of the CTS and Manville TRRS data points and regression curve for the 76.2 mm (3 in.) glass fiber reference specimens.*

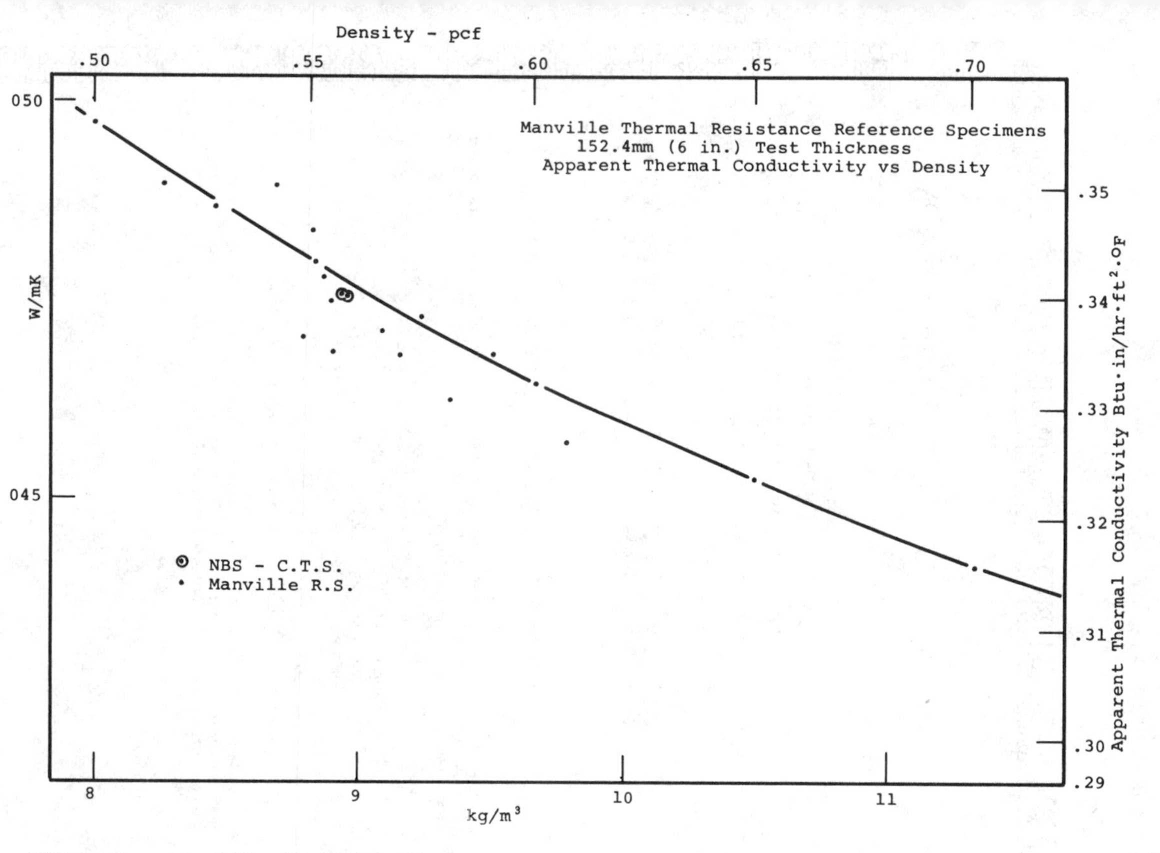

FIG. 5—*Plot of the CTS and Manville TRRS data points and regression curve for the 152.4 mm (6 in.) glass fiber reference specimens.*

boxes. In order to stress the importance of these specimens and their assigned thermal resistance values, a Certificate of Calibration was submitted with instructions on how to proceed with the apparatus calibration. The front and back faces of a facsimile of the calibration certificate are shown in Figs. 6 and 7.

Certificate No. 3035
Date: April 28 '82

MANVILLE CORPORATION
DENVER, CO.
RESEARCH & DEVELOPMENT CENTER
Heat Transfer Technology Laboratory

Calibration Certificate

THERMAL RESISTANCE REFERENCE SPECIMEN

The Thermal Resistance of the specimen identified on this certificate was established by repeated testing of the specimen in an apparatus calibrated with Certified Transfer Specimens (C.T.S.) measured by the National Bureau of Standards, Gaithersburg, MD.

Specimen Number 81-3-35

Thermal Resistance 9.259 F•ft²•hr/Btu, at test thickness
Apparent Thermal Conductivity 0.3240 Btu•in/hr•ft²•°F ;σ 0.00054
Mean Temperature: 75°F; Temperature Difference 50 ± 0.5°F;
Hot Face: 100 ± 0.5°F; Cold Face 50 ± 0. 5°F;
Weight of Specimen; 260.9 grams.
Nominal Density; 0.575 pcf at test thickness; 3.000 inches

The apparent thermal conductivity of this specimen is certified to be accurate within the stated limits of error, σ. This σ is related to the measurements performed by the R/D Center on this specimen and the C.T.S. of identical thickness. A further limitation as to its accuracy is established by the limits of accuracy assigned to the C.T.S. by the National Bureau of Standards; see reverse side.

Research Associate
Heat Transfer Technology

Research Technologist
Calibrations and Standards

FIG. 6—*Reproduction of the Calibration Certificate* (front face) *issued with each individual Manville Thermal Resistance Specimen.*

National Bureau of Standards
Certified Transfer Specimens
Calibration Limitations

Reference: Purchase Order No. Y27-28888-0 dated December 19, 1980

Material:

Specimens were taken from a stock of material obtained by the National Bureau of Standards for providing calibrated thermal resistance reference specimens. The specimen is a batt made of fine-fibered glass and phenolic binder. The specimen is easily compressed, hence should be handled carefully to avoid damage, and for calibration purposes should be used at a thickness close to that indicated under Test Results.

Procedure:

The average thermal resistance of the specimen was measured in a 1000-mm diameter line-heat-source guarded-hot-plate apparatus developed at the National Bureau of Standards conforming with the principles of Standard Method of Test ASTM C177-76. The specimens were dried in an oven at 93°C ± 3°C for at least 4 hours, and cooled to room temperature prior to the measurement. Measurements were made at a specimen mean temperature of approximately 24°C.

The thermal resistance value was determined for that portion of the specimen within the meter area (central 406.4-mm diameter region) of the NBS-GHP-1000. The uncertainty in this thermal resistance value is estimated to be not more than ± 2 percent; this uncertainty includes apparatus systematic error and apparatus repeatability.

The thermal resistance of the specimen has been calculated by correlating thermal resistance to bulk density determinations made on 1) that portion of the specimen corresponding to the meter area of the NBS-GHP-1000 and 2) that portion of the specimen corresponding to the meter area of the user's apparatus**. The overall uncertainty is estimated to be ± "X" percent;* it includes the NBS apparatus uncertainty reported above and the material variability.

Neither the contents of this report nor the fact that the tests were made at the National Bureau of Standards shall be used for advertising or promotional purposes.

25.35 mm (1 inch) specimens; ± 2½%
76.15 mm (3 inch) specimens; ± 2¼-2¾%
152.35 mm (6 inch) specimens; ± 2¼%

**10 × 10 inches for the R-Matic Apparatus.

Note: The above statements are excerpt from the C.T.S. report provided by the National Bureau of Standards.

FIG. 7—*The reverse side of thermal resistance specimen calibration certificate, giving the details of the calibration limitations ascribed by NBS to the CTS.*

Thus, each test location received reference values for calibration of their equipment for three thicknesses. One for the 25.4-mm (1 in.)-thick specimen, two for the 76.2-mm (3 in.)-thick specimens, and one for the 152.4-mm (6 in.)-thick specimen, made up from the two pieces of the 76.2-mm specimens. As stated on each certificate, the values were certified to be referenced to the CTS provided by NBS.

Test Equipment Calibration

With the specimens, each test location received detailed instructions concerning the procedure to be followed for the calibration of their equipment, the development of working reference specimens, and procedures aimed at the maintenance of accuracy of thermal measurements.

The new HFM calibration constant was determined on the basis of random testing of the TRRS. Each specimen was tested four times. These test also provided the confidence limits for the individual apparatus. The need for a single constant for all conditions of test or one for each test thickness was established by the Quality Assurance Department for each test location depending on the characteristics of equipment and the testing requirements. A detailed discussion of this subject as it applied to a Manville plant in Canada is presented in a technical paper by Newton et al [*6*].

Working Reference Specimens

Each plant was instructed to develop their own "working reference specimens" in order to minimize any damage to the TRRS, which are reserved for final verification of the equipment calibration. These working reference specimens were tested four times in order to determine their average thermal resistance value and the standard deviation applicable to each specimen. The values obtained during these tests are used to establish the level of acceptance for any further checks of the equipment through the maintenance of a control chart (company test method) [*6*].

The locally calibrated specimens are used routinely (at least three times a week) to check the equipment performance. The results of these tests are recorded on a control chart for each specimen to observe any deviation in performance and to determine if any change has occurred in the standard deviation as the result of equipment malfunction, improvement, or deficiencies in procedures. Standard test procedures established by the Quality Assurance Department require that corrective measures be taken if certain conditions occur. The recommended procedures for limits of rejection or mandatory action are based on the *ASTM Manual on Presentation of Data and Control Chart Analysis, STP 15D.* If and when a specimen shows signs of damage, mechanical or other, a new specimen is calibrated, and a new control chart is started.

Conclusions

A company wide calibration program has been established. The test equipment provides data which can be referenced to tests performed by NBS.

A program was instituted to ensure continued maintenance of the equipment performance through the institution of a periodic verification of the measurements and maintenance of control charts.

As the result of the implementation of this program the FTC requirements can be fulfilled, thereby insuring that the building insulation products provide the stated insulation performance.

References

[*1*] "Staff Compliance Guidelines for the Federal Trade Commission Rule; Labeling and Advertising of Home Insulation," *Federal Register*, Vol. 45, No. 203. 17 Oct. 1980.

[*2*] Siu, M. C. I. in *Thermal Insulation Performance, ASTM STP 718*, D. L. McElroy and R. P. Tye, Eds., American Society for Testing and Materials, Philadelphia, 1980, pp. 343-360.

[*3*] Pelanne, C. M. in *Proceedings*, 17th International Thermal Conductivity Conference, Gaithersburg, MD, 15-18 June 1981, pp. 763-776.

[*4*] Albers, M. A. and Pelanne, C. M. in *Proceedings*, 17th International Thermal Conductivity Conference, Gaithersburg, MD, 15-18 June 1981, pp. 471-482.

[*5*] Rennex, B., "Low Density Thermal Insulation Calibrated Transfer Samples—A Description and A Discussion of the Material Variability," NBSIR 82-2538, National Bureau of Standards, June 1982.

[*6*] Newton, W. S., Pelanne, C. M., and Bomberg, M. in *Proceedings*, 18th International Thermal Conductivity Conference, 3-5 Oct. 1983, Rapid City, SD, pp. 357-366.

Donald C. Larson[1]

Field Measurements of Steady-State Thermal Transfer Properties of Insulation Systems

REFERENCE: Larson, D. C., **"Field Measurements of Steady-State Thermal Transfer Properties of Insulation Systems,"** *Guarded Hot Plate and Heat Flow Meter Methodology, ASTM STP 879*, C. J. Shirtliffe and R. P. Tye, Eds., American Society for Testing and Materials, Philadelphia, 1985, pp. 206–219.

ABSTRACT: A technique to assess the thermal performance of building envelope systems in the field has been developed. The thermal resistance of these systems is measured with a 1.22-m-square temperature-controlled test plate to which heat flow sensors and thermocouples are attached. The plate is placed on top of an insulated roof system or against an insulated wall system, and thermocouples are attached to the opposite surfaces of the envelope systems. Measurements can be made without disturbing the structure of the envelope systems by using test plate temperatures approximating normal outside surface temperatures. The test plate is either heated or cooled and temperatures and heat flows are measured after equilibrium is achieved. The method provides reliable thermal resistance values for envelope components which are laterally uniform and useful thermal transfer property information for more complex envelope systems. The application of the method to a single-ply expanded polystyrene insulated roof, sprayed-on polyurethane roof insulation systems on both metal and concrete decks, and an insulated metal-stud wall system is described.

KEY WORDS: insulation systems, building envelope systems, thermal resistance, field measurements, heat flow sensors

Standard test methods have been developed to determine the steady-state thermal transmission properties of insulation specimens. The ASTM Test for Steady-State Thermal Transmission Properties by Means of the Guarded Hot Plate (C 177) and the ASTM Test for Steady-State Thermal Transmission Properties by Means of the Heat Flow Meter (C 518) are used to determine the thermal resistance of homogeneous specimens of uniform thickness, and the ASTM Test for Thermal Conductance and Transmittance of Built-Up Sec-

[1]Professor of Physics and Atmospheric Science, Center for Insulation Technology, Drexel University, Philadelphia, PA 19104.

tions by Means of the Guarded Hot Box (C 236) is used to determine the thermal properties of larger nonuniform specimens representative of actual building envelope components. All of these tests are essentially laboratory tests and are not used for field evaluation of insulation systems. It is important, however, to develop tests which could be modifications of the standard tests, for such field evaluations. Building envelope components can be designed or constructed improperly and are subjected to such environmental stresses as freeze-thaw cycles, high humidity, and atmospheric pollution. It is important to make *in situ* evaluations of the thermal performance of such building envelope components to determine both the relevance of the laboratory measurements in assessing the initial state of the components and the ability of the envelope components to maintain their initial-state properties.

Many different approaches have been used to evaluate the thermal properties of building envelope components in the field. Perhaps the most elaborate scheme is that employed by investigators at the Lawrence Berkeley Laboratories [*1*]. They developed an envelope thermal testing unit which consists of two blankets which are placed against the opposite sides of a wall. Each blanket contains two heating sections which allow the heat flux into each side of the wall to be controlled separately, and the blankets are slightly flexible so that they can conform to slight irregularities in the wall surfaces. The device is portable, and the heat flows rather than the temperatures are the independent variables. The use of a "pink-noise" driving function for the heat flows allows simplified thermal parameters, which characterize the dynamic thermal performance of the wall, to be determined. Another device which is also similar to a guarded hot box, but which is designed for field measurements, has been developed by the Building Research Division of the National Research Council of Canada [*2*]. This device is a calorimeter with a metering area of 1.2 by 2.1 m which is placed on the inside of a wall so that the conditions at the exterior of the wall are not altered during the measurement. This device has the advantage of being usable with laterally nonuniform walls and the disadvantages of being rather bulky and requiring careful control of the interior temperatures during measurements. Field measurements of building envelope components can be also made by attaching temperature and heat flow sensors on the inside and outside surfaces. Long-term average heat flows and temperature differences allow average thermal resistances of the building components to be determined. This method has been used with insulated walls [*3,4*] and roofs [*5,6*]. The testing is done with the building component in its actual environment, but care must be employed in using heat flow sensors due to their sensitivity to wind and other environmental conditions and to the method of their surface attachment.

The field testing approach described here employs heat flow sensors semipermanently attached to a temperature-controlled 1.22-m (4-ft) square test plate. The test plate is portable and is used for *in situ* thermal resistance measurements of building envelope components. Typically the plate is placed against the outside surface of the building or building component, is heated or cooled to an equilibrium temperature either above or below the inner surface temperature, and the inside and outside surface temperatures and the heat flow are measured. Usually the test plate is heated in the summer and cooled in the winter so that the measurement may be made without disturbing the structure of the building component. The test plate is covered with an insulated lid so that variations in the outside temperature during a test will not cause significant edge heat flows. The test plate is small enough to be portable, but it uses multiple heat-flow sensors so that it can measure the heat transfer through a building component over a reasonably large area. When the envelope insulation system is laterally nonuniform, the test plate is moved from one location to another to ensure that representative areas of the envelope component are sampled and realistic average thermal parameters are obtained. The test plate provides constant temperature boundary conditions on the outer surface of the building component, but the conditions on the inside are not subject to the same controls. In practice, however, the inside air temperature is sufficiently stable so that a reliable value for the thermal resistance is obtained. Except for this lack of control of the inside surface temperature, the test plate method is similar to the laboratory scale ASTM Test for Steady-State Thermal Transmission Properties by Means of the Heat Flow Meter (C 518). Such laboratory tests, however, employ relatively small homogeneous specimens, whereas a field test method must be applicable to real envelope structures where one-dimensional heat flow cannot be assured.

The test plate method was used initially to measure the thermal resistance of constant thickness, laterally uniform 1.22-m (4-ft) square insulated roof systems enclosed in insulated modules. The calibration procedure and initial results have been published [7]. In this paper the application of the method to more complex building envelope systems is described.

Experimental

Test Plate

The thermal resistances of building envelope components are determined by measuring heat flow rates and temperature differences under steady-state conditions. The primary experimental device is a 1.22-m (4-ft) square, 0.95-cm (0.38-in.) thick aluminum test plate which is shown in cross section in Fig. 1. Three heat flow sensors spaced 30.5 cm (12 in.) apart are attached to the bottom of the plate with heat sink compound. Two sheets of an elastomeric membrane of 1.14-mm (0.045-in.) ethylene propylene diene monomer rubber (EPDM) serve as a guard for the heat flow sensors; they are cut out and glued to the plate. A third EPDM sheet is glued to the plate to cover the heat flow sensors; this sheet is attached to the heat flow sensors with heat sink compound. The three heat flow sensors were purchased from two different

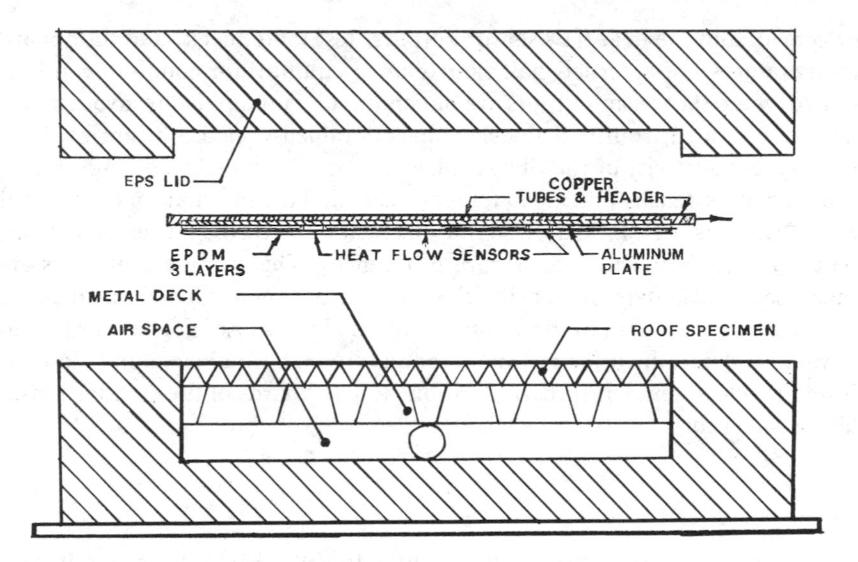

FIG. 1—*Roof module test configuration.*

manufacturers, are 5.1 cm (2 in.) and 7.0 cm (2.75 in.) square, and all have integral Type T thermocouple junctions. The top of the test plate is fitted with copper tubes and headers for the heat transfer fluid. A uniform fluid temperature is provided by a programmable thermostatically controlled bath and circulator. During a test, the test plate temperature will vary laterally by less than 0.2°C except where the envelope component being tested has thermal shorts where temperature differences as large as 2.0°C have been observed.

Calibration

The heat flow sensors are calibrated by testing specimens of known thermal resistance under conditions which approximate actual test conditions. The calibration procedure is therefore similar to that employed in an ASTM heat flow meter test (C 518). Five fibrous glass boards with phenolic binder were obtained from the National Bureau of Standards; the boards are 1.22 m (4 ft) square, 2.5 cm (1 in.) thick, and have a nominal density of 160 kg/m^3 (10 lb/ft^3). One to five boards of this material are sealed into an insulated module over a metal deck; this calibration module is similar to the test module shown in Fig. 1. Laboratory air is circulated underneath the metal deck of the module. Type T thermocouples are used to measure the metal deck temperature and the temperatures between the boards. One layer of EPDM is placed over the boards for protection, and the test plate is placed on top and then covered with an insulated lid. The thicknesses of the boards are measured with a caliper with the test plate in place. The test plate is either heated or cooled, and the heat flow sensor outputs, test plate, deck, and interboard temperatures are recorded half hourly until equilibrium conditions are attained where the sensor outputs are not changing monotonically; average values are obtained from half hourly values obtained over a 3 h interval. The thermal conductivity of the fibrous glass board is assumed to be known as a function of its density and mean temperature, and the thermal conductivity of the EPDM has been measured so that the heat flows through each board may be calculated from the temperature differences. The average heat flows are then used to calculate the sensitivities of the heat flow sensors defined as the ratios of the heat flows to the voltage outputs of the sensors. The sensitivities have been determined for test plate temperatures both above and below the inside envelope temperatures and for one to five boards of the standard insulation material.

Facilities and Procedures

Testing has been performed on modular 1.22-m (4-ft) square roof insulation systems, a sprayed-on foam insulation system on a concrete deck, and on an insulated metal stud wall. The roof module system is shown schematically in Fig. 1. The 1.83-m (6-ft) square module with a 1.22-m (4-ft) square cavity was first constructed of EPS block density 24 kg/m^3 (1.5 lb/ft^3). Fluted galvanized decking supported at its perimeter with a galvanized metal frame was then placed into the cavity. The roof specimens were then prepared in the same manner as if the metal decking were a part of a commercial flat roof. An EPS roof specimen was constructed by first placing a 1.22-m (4-ft) square EPS board directly onto the deck and then covering it with an elastomeric membrane of 1.14-mm (0.045-in.) thick EPDM. A polyurethane roof specimen was prepared by taping over the fluted deck and then spraying successively with a polyurethane foam, an elastomeric coating, and an aluminized coating. The cavity beneath the fluted deck forms a flow passage through which a controlled flow of air is circulated from the adjacent laboratory. The undersides of the modular roof insulation systems are therefore subjected to normal building interior conditions, whereas the outer surfaces of the insulation systems are exposed to the outdoor environment. Type T thermocouple sensors are attached to the metal deck. The concrete deck of the laboratory was also insulated with a sprayed-on polyurethane foam system with the same nominal thickness as the modular system. The laboratory roof is 5.7 by 12.8 m (19 by 42 ft), and the concrete deck is 10.8 cm (4.3 in.) thick; a representative cross section is shown in Fig. 2. Also shown is a cross section of an insulated interior wall of the insulation laboratory. The wall has vertical galvanized steel studs 0.51 mm (0.02 in.) thick with a 40.6 cm (16 in.) separation, 8.9 cm (3.5 in.) of aluminum foil faced fiber glass blanket insulation, and 1.3 cm (0.5 in.) gypsum board sheathing. Type T thermocouple sensors

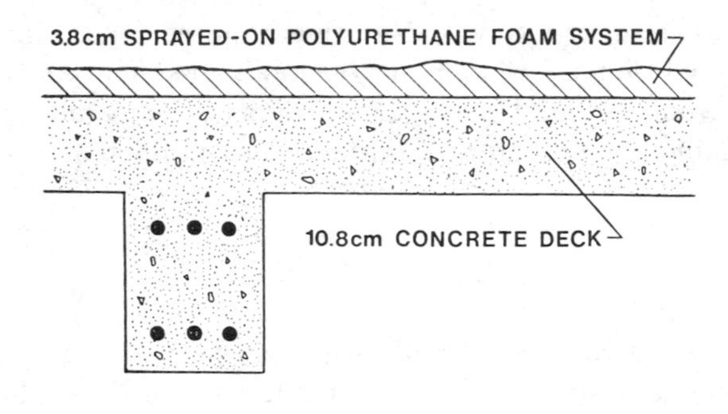

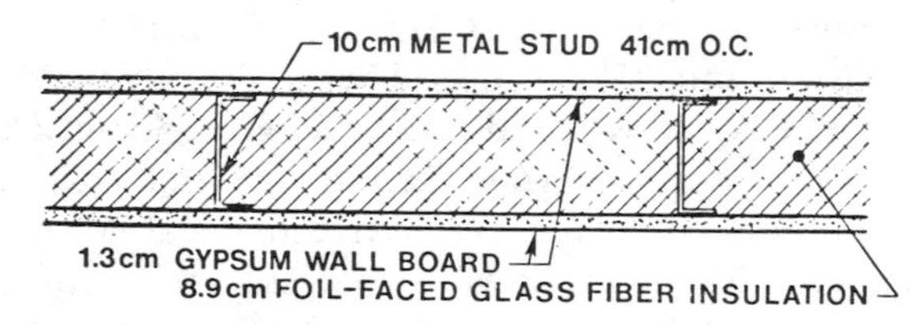

FIG. 2—*Cross sections of roof and wall insulation systems.*

were attached under the concrete deck at the test locations and also to one side of the interior metal-stud wall.

The test plate was used to determine the thermal resistance of the insulated roof specimens, the insulated laboratory roof, and the laboratory partition wall. For the insulated roof systems the test plate was simply placed on the roof system and covered with an insulated lid as shown in Fig. 1. The sprayed-on insulation had an irregular surface so it was necessary to place foam rubber around the edges of the plate to prevent air from leaking under the plate during a test. The test procedure was the same as that employed during the calibration, and the thermal resistance of the systems was calculated from the plate and below deck temperatures and the heat flow sensor outputs. The laboratory roof was tested at three separate locations, and the insulation thickness was measured at the location of the heat flow sensors at all three locations. The thermal resistance of the insulated wall was measured by placing the test plate vertically against the wall and then covering it with the insulated lid. The test plate was oriented with the heat flow sensors horizontal so that the three sensors were aligned perpendicular to the metal studs. The thermal resistance at three locations along the wall was determined from the wall and plate temperatures and the heat flow sensor outputs.

Heat flows and surface temperatures were recorded at half-hourly intervals. After equilibrium was achieved, where the sensor outputs were not changing monotonically, average heat flow and temperature values were obtained over a three-hour interval. The time required for equilibrium to be established varied from a few hours for the modular systems to 10 to 12 h for the laboratory roof system. The longer test runs were generally performed at night when ambient temperature variations were minimal and solar loading was absent.

Results

Three calibration runs were made with two layers of the fibrous glass boards, and one calibration run each was made with one, three, four, and five layers. The test plate was heated for these runs, and the average insulation temperature ranged between 40 and 44°C (103 and 111°F). The sensitivities of heat flow sensors 1, 2, and 3 were 1.31, 1.13, and 5.16 W/m^2 mV, respectively, for two layers of the fibrous glass board. The sensitivities were about 4% higher for one layer and 4, 7, and 10% lower for 3, 4, and 5 layers, respectively. The decrease in heat flow sensor sensitivity with increasing calibration sample thickness is due to the increase in fringing heat flow with increasing thickness. Further calibration runs will be necessary for more accurate sensitivity values for these insulation thicknesses. One below ambient (mean temperature 1°C) calibration run with two fibrous glass boards gave sensitivities only 2% above those previously quoted, but again further calibrations are required.

The results obtained in testing the modular insulation systems are given in Table 1. The first three test runs shown were made in the laboratory with an expanded polystrene (EPS) board with density 15 kg/m^3 (0.94 lb/ft^3) and thickness 6.20 cm (2.44 in.). The next three test runs were made with the same EPS board in an outdoor test module. The heat flow values were obtained by using sensitivities 1.8% lower than those obtained for the two fibrous glass board calibrations to account for the different thicknesses employed in the calibration and test runs. The mean insulation system temperatures varied from 38 to 44°C (100 to 111°F) and were close to the temperatures employed in the calibration runs. The thermal resistance values were calculated from the equation $R = \Delta T/\bar{Q}$, and no systematic variation in the R-values was observed with indoor and outdoor measurements, or over the one-year period of outdoor testing. The final three test runs shown in Table 1 were made with the sprayed-on polyurethane system. The thickness of the insulation system was measured at the location of each of the three heat flow sensors and found to be 40.5, 43.0, and 40.5 mm for an average value of 41.3 mm (1.63 in.). The heat flow values were obtained using sensitivity values 1.5% higher than the values obtained for two layers of fibrous glass material due to the lower average thickness of the polyurethane insulation system. The mean insulation temperatures were in the same range as those employed with the EPS roof modules.

The thermal resistance of the insulated laboratory roof was measured at

TABLE 1—*Module test results.*

Date	Description	Temperature Difference, °C	Heat Flow, W/m²				Thermal Resistence, m² K/W (ft² h °F/Btu)		Mean Temperature, °C
		ΔT	Q_1	Q_2	Q_3	$\bar{Q}$	R		T_M
6/24/81	EPDM-EPS, CM	22.9	16.3	16.4	16.2	16.3	1.40	(7.95)	42.8
7/20/81	EPDM-EPS, CM	26.0	19.1	18.8	18.7	18.9	1.38	(7.81)	43.8
8/17/81	EPDM-EPS, CM	18.1	13.0	12.9	13.0	13.0	1.39	(7.91)	38.1
8/19/81	EPDM-EPS, TM	17.6	12.3	12.4	12.6	12.5	1.41	(8.01)	37.9
9/21/81	EPDM-EPS, TM	22.1	16.2	15.3	15.8	15.8	1.40	(7.94)	40.6
11/11/82	EPDM-EPS, TM	27.2	20.4	19.2	19.5	19.7	1.38	(7.81)	38.6
9/26/81	U, TM	14.6	8.64	8.74	8.07	8.48	1.72	(9.78)	45.0
9/29/81	U, TM	26.8	14.7	15.0	14.0	14.5	1.85	(10.5)	38.6
11/12/82	U, TM	25.8	15.4	15.1	14.9	15.1	1.71	(9.69)	39.2

NOTES—EPDM = elastomeric membrane made of ethylene propylene diene monomer rubber.
EPS = expanded polystyrene.
U = sprayed-on polyurethane foam insulation system.
CM = calibration module.
TM = test module.

three locations and the experimental results are given in Table 2. The resistance values at each test location were calculated from average temperatures and average values of the heat flows; heat flow values were calculated using the same heat flow sensor sensitivities used for the comparable test module. At location No. 1 the thermal resistance was the highest, and the insulation thickness was also highest in this region; the average insulation thickness at the location of the heat flow sensors was equal to the nominal 38 mm (1.5 in.) thickness. The insulation thickness was lower at the other locations and correspondingly the measured thermal resistances were also lower at these locations. The thermal resistance and average insulation thickness were both about 30% lower at test location No. 3 than at test location No. 1. There was also a correlation between the heat flows recorded by individual heat flow sensors and the insulation thickness measured below each heat flow sensor; the highest heat flow (24.3 W/m^2 was recorded at the point at test location No. 3 where minimum insulation thickness (17 mm) was recorded. There was a surface depression at this point, and heat flow sensor No. 3 was deliberately located over this depression. The thermal resistance and average insulation thickness values obtained at location No. 1 over the concrete deck are close to the thermal resistance and average thickness values obtained in the test module with the steel deck.

The thermal resistance of the insulated partition wall was also measured at three locations, and the results are given in Table 3. The resistance values were again calculated from the average temperatures and the average heat flows. The minimum center-to-center heat flow sensor to stud flange separations are also given on the table. In test location No. 3 heat flow sensor No. 2 was almost directly over the stud flange, while heat flow sensor No. 1 was midway between the studs. The heat flow recorded by sensor No. 2 was 4.6 times greater than the heat flow recorded by sensor No. 1. The thermal resistance had a minimum value at this location. In test location No. 2 none of the heat flow sensors were closer than 41 mm from the stud flange, and in this location the measured thermal resistance was highest. The average value of the thermal resistance was 1.17 m^{2}K/W (6.62 ft^2 h °F/Btu). In the wall test the heat flow sensor sensitivities were assumed to be the same as those obtained in the two-layer fibrous glass board calibration tests.

Discussion

The application of the test plate method is relatively straightforward in determining the thermal transfer properties of homogeneous or laterally uniform insulation systems. The method is then similar to a laboratory heat flow meter test where standard insulation materials are used for calibration. With both methods it is important that the specimen thickness, mean temperature, and testing conditions be the same when test runs are made with both calibration material and the test material. If a test method is to be used to evaluate

TABLE 2—*Roof test results.*

Test Location	Temperature Difference, °C	Heat Flow, W/m²				Thickness, mm				Thermal Resistance, m² K/W (ft² h °F/Btu)		Mean Temperature, °C
	ΔT	Q_1	Q_2	Q_3	$\bar{Q}$	t_1	t_2	t_3	$\bar{t}$	R		T_M
1	29.9	16.4	17.8	17.0	17.1	42	34	39	38	1.75	(9.93)	41.9
2	27.4	18.7	17.3	17.9	18.0	25	32	29	29	1.53	(8.66)	43.9
3	26.9	17.7[a]	20.5	24.3	22.4[a]	33	25	17	25	1.20	(6.82)	43.7

[a] Heat flow sensor No. 1 was located over a beam; the thermal resistance was calculated using sensors Nos. 2 and 3.

TABLE 3—*Wall test results.*

Test	Temperature Difference, °C	Heat Flow, W/m²				HFS to Stud Distance, mm			Thermal Resistance, m² K/W (ft² h °F/Btu)		Mean Temperature, °C
	ΔT	Q_1	Q_2	Q_3	$\bar{Q}$	d_1	d_2	d_3	R		T_M
1	24.5	8.0	12.6	41.6	20.8	89	117	16	1.18	(6.69)	37.7
2	23.8	7.8	24.8	25.1	19.2	159	41	64	1.24	(7.04)	37.3
3	26.9	9.9	45.3	19.9	25.0	197	3	102	1.08	(6.11)	39.8

insulation systems in the field, it becomes difficult to control the testing conditions and to obtain suitable calibration standards. At the same time it is important to develop field testing methods; laboratory conditions are not the same as field conditions so their use in evaluating insulation systems is limited. The methodology employed in our testing program is to evaluate (1) uniform well characterized insulation systems in a laboratory environment, (2) the same systems in the field, (3) similar systems in the field, and (4) more complex insulation systems in the field using, where necessary, numerical calculations to relate the measurements to meaningful performance parameters. The results just given illustrate this approach. The first three tests with EPS insulation were performed in the laboratory in the same configuration used in the calibrations. The insulation board was then placed in an outdoor test module which is essentially the same as the indoor calibration module; similar thermal resistance values were obtained in both cases as shown in Table 1. A similar test module was then used to evaluate a sprayed-on polyurethane foam system which at the same time was applied to the laboratory roof. Field tests on the laboratory roof with a concrete deck can then be compared to the smaller scale modular tests of the same insulation system. This comparison is illustrated in Tables 1 and 2. Finally, the test method was used to evaluate the thermal transfer properties of a wall with substantial lateral property variations.

The sprayed-on polyurethane system has a variable thickness so that lateral temperature differences will develop along the outer surface of the insulation system in the field. The temperature-controlled test plate will bring the outer surface temperature of the system to an approximately uniform temperature so that the boundary conditions of the insulation system are different during testing than in the field. Except for test location No. 3, however, the thickness varied by less than ±12% from the mean thickness, and the variation of the heat flow recorded by the three test plate sensors was even less. For this reason the thermal resistance values obtained with the isothermal test plate can be considered to be reliable measures of the thermal transfer properties of this type of insulation system. Even at test location No. 3 where large thickness variations were observed, the measured thermal resistance correlated well with average thickness; the thermal resistance was 31% lower than the thermal resistance at test location No. 1, and the average insulation thickness was 34% lower at this location than the insulation thickness at location No. 1. A variation in average insulation thickness would not be readily detectible with other testing techniques so that the test plate method served a useful function in the thermal evaluation of this type of insulation system.

The differences in the heat flows recorded by the individual heat flow sensors were relatively small in the laboratory roof tests (Table 2), but were larger than the differences observed with the insulation systems of constant thickness as shown in Table 1. The individual heat flow sensor outputs, however, varied widely in the insulated wall tests. As seen in Table 3, the heat flow sensor outputs are related to the sensor locations relative to the steel stud locations. When a heat flow sensor is close to a stud a relatively large heat flow is recorded. In spite of this variation the thermal resistance values obtained at all three locations were not greatly different because these values were based on the average heat flows from the three sensors.

Nonuniform heat flow through an insulated wall will be accompanied by a nonuniform surface temperature. The test plate, however, causes one surface to remain at a relatively uniform temperature so that the testing conditions are different than the field conditions. For this reason the R-values obtained with this method should be labelled effective R-values. Further research involving both experiments and numerical heat transfer calculations will be required to relate these effective R-values to the generally accepted R-values obtained with a guarded hot box. The effective R-values shown in Table 3, however, provide a clear indication of thermal bridging; the thermal resistance of such a system is much lower than the value calculated by assuming one-dimensional heat flow. If we use values of $R_I = 1.94$ m^2 K/W (11 ft^2 h °F/Btu) for the fiber glass blanket insulation, $R_s = 2.26 \times 10^{-3}$ m^2 K/W (0.0128 ft^2 h °F/Btu for the steel stud ($k = 45$ W/m K), and $R_w = 7.93 \times 10^{-2}$ m^2 K/W (0.45 ft^2 h °F/Btu) for the wall board, the wall overall thermal resistance would be given by the parallel heat-flow-path method as

$$1/R_T = A_I/(2R_w + R_I) + A_s/(2R_w + R_s) \quad (1)$$

where A_I and A_s are the fractional cross-sectional areas of the insulation and stud, respectively. The stud width and spacing were measured to be 0.51 and 406 mm (0.02 and 16 in.), respectively, so that $A_s = 0.00125$, $A_I = 1 - A_s$, and $R_T = 2.07$ m^2 K/W (11.7 ft^2 h °F/Btu). This value is close to the value obtained by summing the insulation and wall board resistance and is of course unrealistic in that it neglects the transverse heat flow in the flange of the stud as well as in the wall board.

The Zone Method has been proposed as an approximation which provides a more realistic thermal resistance value for an insulation system that contains metal or other highly conductive material [*8*]. In this method the overall thermal resistance would be given by

$$1/R_T = A_a/R_a + A_b/R_b \quad (2)$$

where A_a is the fractional area of the zone which contains the metal and $A_b = 1 - A_a$ is the fractional area of the zone which contains no metal. The recommended value for A_a would be calculated as $(w_f + 2d_w)/w$ where $w_f = 31.8$ mm (1.25 in.) is the width of the flange of the stud, $d_w = 12.7$ mm (0.5 in.) is the thickness of the wallboard, and $w = 406$ mm (16 in.) is the stud spacing. The fractional areas of Zones A and B are therefore $A_a = 0.141$ and

$A_b = 0.859$, respectively. The thermal resistance of Zone B is simply $R_b = 2R_w + R_I$, whereas the thermal resistance of Zone A is given by

$$R_a = 2R_w + R_s R_I A_a/(R_s A_I + R_I A_s) \quad (3)$$

where A_I is now given by $A_a - A_s$. Using the values for the parameters given previously, the overall thermal resistance of the wall is calculated from Eq 2 to be 1.29 m K/W (7.32 ft h °F/Btu), and R_a and R_b are 0.384 and 2.10 m K/W, respectively. In this approximation the metal stud lowers the thermal resistance of the entire Zone A, and the resulting thermal resistance of the wall is much closer to the experimental result than the value obtained using the parallel heat-flow-path method.

The Zone Method appears to provide a useful approximate calculation of the wall overall thermal resistance. It does not provide, however, insight into the physical mechanism which cause the experimental effective thermal resistance to be 43% lower than the value calculated assuming one-dimensional heat flow. The low experimental thermal resistance is due to the lateral heat flow that occurs in the wall board, in the stud flanges, and also in the aluminum foil facing on the insulaton. Although the aluminum foil is only 0.036 mm (0.0014 in.) thick, its thermal conductance is about four times greater than the gypsum wall board which is 12.7 mm (0.5 in.) thick. The air-wall surface temperature differences will also be greater in the vicinity of the studs causing increased heat transmission in these regions. A simple two-dimensional heat flow model which allows lateral heat flow in the wallboards, stud flanges, and insulation foil facing has been developed by the author which yields an overall thermal resistance value for the metal stud wall of 1.28 m^2 K/W (7.3 ft^2 h °F/Btu). This value is quite close to the value obtained with the empirical Zone Method. All of the calculated values are based upon material properties measured at 23.9°C (75°F). Since the mean test temperature was 38.3°C (101°F), the insulation thermal resistance R_I was lower than the value cited previously. It appears therefore that the experimental values are quite close to the values calculated with either the Zone Method or a method which considers two-dimensional heat flow.

The test plate method has been used to measure the thermal resistance of many different types of insulated roof systems and also the effective thermal resistance of a metal stud wall. The method has been found to be convenient, readily adaptable to field measurements, and capable of providing accurate thermal resistance values for laterally uniform insulation systems. The application of the method to laterally nonuniform insulation systems will require further study involving numerical calculations and perhaps guarded hot box experiments.

Acknowledgments

The author would like to thank Robert Zarr for useful discussions, Thomas Edmunds and Jon Larson for their help with experimental measurements, and ARCO Chemical Co. for its financial support.

References

[*1*] Sherman, M. H., Adams, J. W., and Sonderegger, R. C. in *Thermal Insulation, Materials, and Systems for Energy Conservation in the '80s, ASTM STP 789*, American Society for Testing and Materials, Philadelphia, 1982, pp. 355–372.

[*2*] Brown, W. C. and Schuyler, G. D. in *Proceedings*, ASHRAE/DOE-ORNL Conference on Thermal Performance of the Exterior Envelopes of Buildings, Kissimmee, FL, Dec. 1979, pp. 262–268.

[*3*] Tao, S. S., Bomberg, M., and Hamilton, J. J. in *Thermal Insulation Performance, ASTM STP 718*, American Society for Testing and Materials, Philadelphia, 1980, pp. 57–76.

[*4*] Flanders, S. N. and Marshall, S. J. in *Transactions*, ASHRAE, Vol. 88, No. 1, 1982, pp. 677–687.

[*5*] Hedlin, C. P., Orr, H. W., and Tao, S. S. in *Thermal Insulation Performance, ASTM STP 718*, American Society for Testing and Materials, Philadelphia, 1980, pp. 307–321.

[*6*] Treado, S. J., "Thermal Resistance Measurements of a Built-Up System," NBSIR 80-2100, Oct. 1980.

[*7*] Larson, D. C. and Corneliussen, R. D. in *Thermal Insulation, Materials, and Systems for Energy Conservation in the '80s, ASTM STP 789*, American Society for Testing and Materials, Philadelphia, 1982, pp. 400–412.

[*8*] ASHRAE Handbook, Fundamentals Volume, 1981 Ed., Chapter 23, American Society of Heating, Refrigerating, and Air-Conditioning Engineers.

Stanley P. Schumann[1]

Surface Temperature Sensor Calibration: *In Situ* Technique

REFERENCE: Schumann, S. P., **"Surface Temperature Sensor Calibration: *In Situ* Technique,"** *Guarded Hot Plate and Heat Flow Meter Methodology, ASTM STP 879*, C. J. Shirtliffe and R. P. Tye, Eds., American Society for Testing and Materials, Philadelphia, 1985, pp. 220–226.

ABSTRACT: The surface temperature dectors on a guarded hot plate can be calibrated in place using differential thermopiles. A ten-junction Type T thermopile was used to calibrate a 1.22-m (4-ft) guarded hot plate using water triple point cells as the reference temperature. The stability and precision of these thermopiles, over a year's span, was on the order of 20 to 30 millidegrees.

KEY WORDS: guarded hot plate, thermopiles, senor calibration, thermopiles

Thermal resistance testing of insulating products demands precise and accurate measurement of surface temperatures. To evaluate the effect of thickness on the measurement of building insulation R-values, Owens Corning designed and built a 1.22-m (4-ft) square guarded hot plate (GHP). At the inception of testing the absolute value of the thickness effect was unknown, but we suspected it to be relatively small. Accumulated errors in temperature difference, thickness measurement, etc., could possibly mask or exceed any deviations in the test data.

During the initial testing stages the need for calibration of the surface temperature measurement system in the GHP became evident to us. The temperature sensors were embedded in the GHP's isothermal plates and thus could not be removed for a careful laboratory recalibration [*1*]. We devised a procedure to calibrate them *in situ* using calibrated transfer sensors. Our procedure had to be accurate, precise, and not mar the GHP surfaces.

Accuracy was necessary for us to fully evaluate the test data, while good precision was needed to give us confidence in the data. In order to have an overall GHP error of less than 1%, we needed to establish error limits on the

[1]Engineer, Technical Center, Owens Corning Fiberglas, Granville, OH 43023.

temperature, power, and thickness measurements to stay within the 1% overall error. If less than one third of the overall error was to come from the temperature difference then the plate surface temperatures needed to be known within 0.1°C each.

Finally, the calibration could not be allowed to damage the plate surfaces as they were expensive ground surfaces. We could not attach calibration sensors by adhesives or by peening them into the plate surfaces [*1*]. This restriction was perhaps the most demanding to meet as accurate measurements are difficult without intimate contact.

Method Theory

In order to measure the surface temperature over the embedded sensor we chose to use ten-junction thermopiles as the measurement sensors, which would effectively average the surface temperature over a 50 to 75 mm area over the sensor. The reference used was the triple point of water which can be maintained in a commercial cell within 0.01°C with an accuracy and precision of better than a millidegree. Thus, our signal from the thermopile would give us the difference between the reference cell and the plate surface temperature. With an accurate reference the plate temperature could be accurately determined.

Thermopiles

Our four thermopiles were constructed using standard grade, Teflon-coated Type T thermocouple wire (Fig. 1). The beads were welded using an inert gas arc welder and coated with a polyurethane varnish, normally used for strain gages, to electrically isolate the beads from ground loops. The thermopiles were heat aged at a temperature of about 62°C for several days to stabilize the thermopiles.

Ten junctions on each end were used so the output electromotive force (EMF) would be ten times greater than a single junction Type T thermocouple. Using voltage measurement instruments with a resolution of 1 μV gave us a resolution of 0.0025°C using the thermopiles. This is a tenfold increase in resolution over a single junction thermocouple.

Support Equipment

We measured the thermopile output using a data logger. The thermopiles were wired to the data logger using isothermal cards to reduce spurious signals caused by thermal EMFs at the connections. To further reduce these spurious signals the thermopile output was measured twice: once with the normal polarity and again with the leads reversed. When the absolute values of these measurements are averaged, any zero offset caused by thermal EMFs

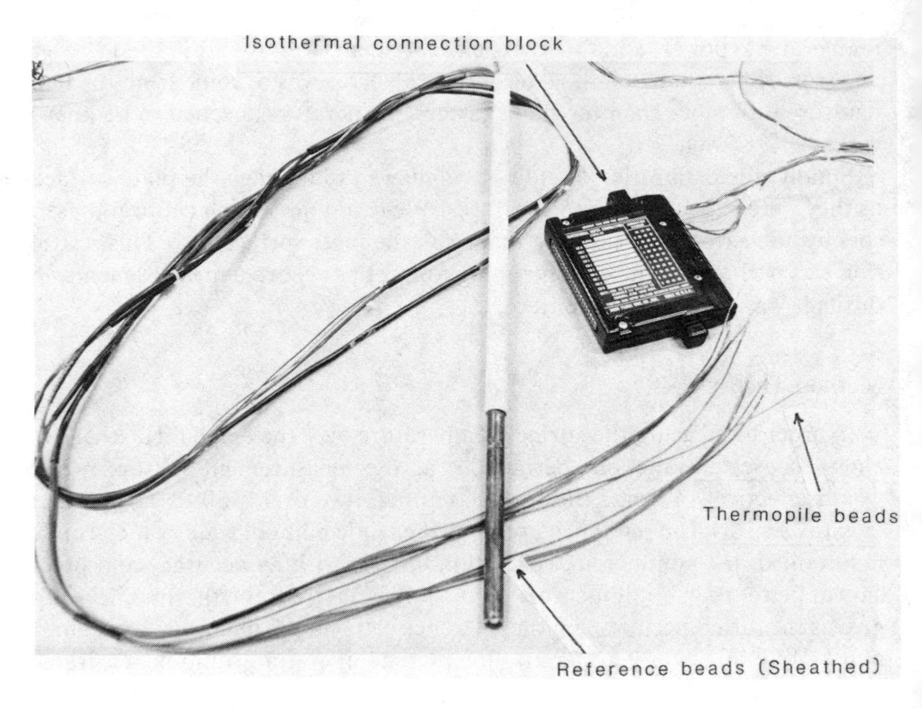

FIG. 1—*Photograph of one thermopile.*

in the connection averages cut. The reversal was a simple wiring change in the isothermal connector.

Another reason we chose the data logger was the interface option available for the data logger. We used a microcomputer to control the data logger, display plate surface temperatures, and store data onto magnetic tape for later analysis.

Although we wanted to use a standard platinum resistance thermometer as our temperature standard the delivery time from the manufacturers were too long. Instead we chose a four wire thermistor standard which was calibrated by the manufacturer traceable to the National Bureau of Standards (NBS). Overall accuracy of this standard was quoted as within 0.01°C.

To measure the probe resistance we devised a constant current source to supply a 10 μA current regardless of the thermistor resistance. Self-heating errors with the low current were kept at a minimum. Two sets of precision resistors (0.01 and 0.02%) monitored the current loop, one set before the probe and one set after the probe. We used 4-wire ohm measurement techniques to measure the resistors in both the normal and reversed EMF modes. Again the absolute values of the measurements were averaged to obtain the resistance.

Four thermopiles were calibrated against the standard thermistor in a specially constructed constant temperature bath. We designed the bath for accurate temperature calibrations in the range from 0 to 100°C. The bath temperature is maintained with a small heater and a precision temperature controller, while a 100-L tank of chilled ethylene glycol/water solution supplies cooling as needed. The chilled tank is maintained around 4°C with a portable cooling unit. Recent tests with a standard platinum resistance thermometer show that the system is able to maintain bath temperature to within ±0.003°C of the desired value over a range from 7 to 46°C.

The thermopiles were calibrated at the following nominal temperatures: 8.9, 10, 11.1, 11.7, 12.2, 12.8, 13.3, 13.9, 33.9, 34.4, 35, 35.6, 36.1, 36.7, 37.8, 38.9, and 44.4°C. These temperatures were chosen to give a reasonable spread around 35 and 12.8°C where the GHP operates and also around 37.8 and 10.0°C where NBS conducts thermal tests.

Twenty individual readings were taken at each nominal temperature and the mean values used for calibration regressions. Typical standard deviations for the thermopiles were on the order of 1 to 2 μV or equivalent to about 0.002°C. We obtained interpolation equations by simple linear regression [*2*] on the thermopile output. The thermopiles were found to be linear, within 0.005°C, over a 3°C range as, shown by Fig. 2. Regression to fit a second

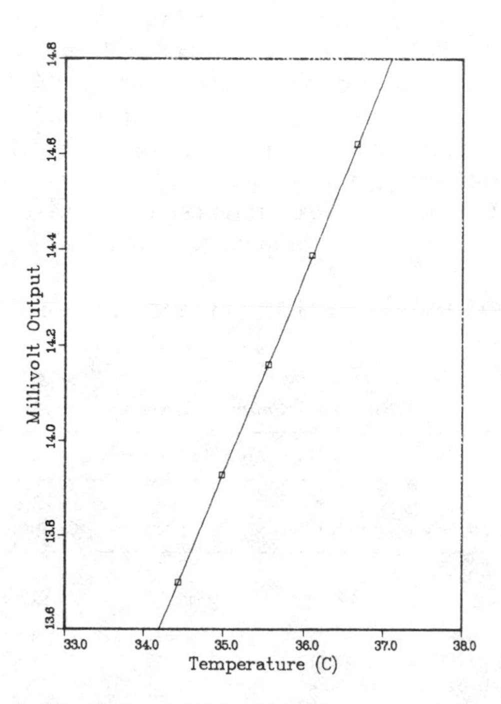

FIG. 2—*Typical calibration curve. Thermopile calibration: millivolts versus degrees Celsius.*

order [2] yielded marginally better results; however, for general use, the linear fit is adequate.

We have used the thermopile technique three times now, and the thermopiles appear to be very stable over time. Table 1 was constructed using voltages generated during the latest calibration applied to the previous calibration equations. The results of these three calibrations are more outstanding considering that our measurement equipment evolved over the calibration series so that different equipment was involved in the first two calibrations.

Calibration of the GHP

With the thermopiles calibrated, we began on the GHP. The thermopiles were sandwiched in polyester film with thermal grease for good contact, and held in place with masking tape (Fig. 3). Fiber glass boards, 25 mm thick, were placed next to each thermopile to provide a firm backing. Between the fiber glass boards, 300 mm of rigid foam was sandwiched. The total test thickness in the GHP was then 350 mm. The large test resistance was chosen to reduce the heat flux through the sample. This was necessary as the resistance temperature device (RTD) sensors in the GHP were embedded slightly below the plate surfaces. Figure 4 shows a side view of the calibration assembly.

Resistance measurements from the GHP RTDs were entered manually into the computer as the GHP data system was not easily interfaced with the calibration package. This allowed us to directly calibrate the GHP RTD's measured resistance with the measured surface temperature and thus calibrate the entire GHP temperature measurement system. Three sets of readings were taken at each temperature setting with 20 scans of all data channels taken each set. The thermopile calibration equations were programmed into the calibration package to allow immediate readout of the measured surface temperature.

We performed all these calibrations while the specimen was at equilibrium,

TABLE 1—*Stability of thermopiles.*

	Thermopile Temperatures, °C		
Output, mV	1st Calibration	2nd Calibration	3rd Calibration
4.953	12.76	12.76	12.75
4.733	12.20	12.19	12.19
13.928	34.96	34.97	34.98
13.700	34.41	34.42	34.43
14.161	35.52	35.53	35.55

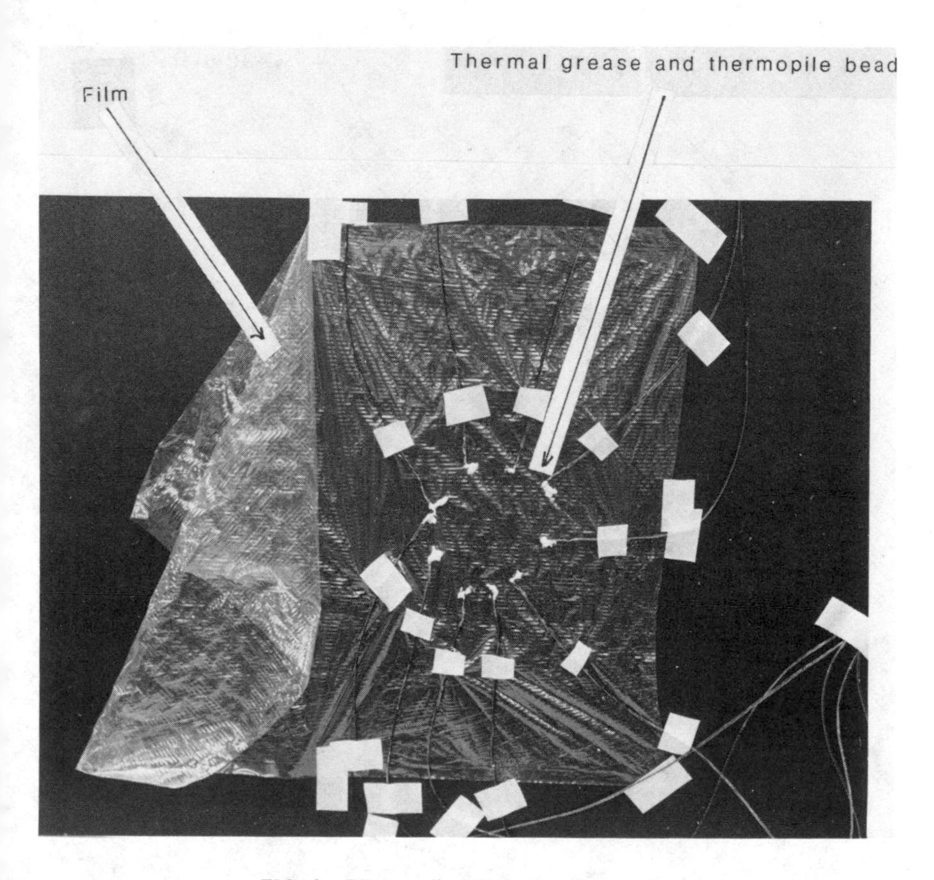

FIG. 3—*Thermopile attached to plate surface.*

which required at least 24 h to occur. To attempt to reduce the equilibrium time we removed the foam and rechecked several temperatures. Differences between the 350 and 50-mm calibrations were −0.06°C on the hot side and 0.01°C on the cold side. Since we considered this excessive, the 300-mm foam boards were replaced and the calibration spot-checked. The two 350-mm calibrations agreed so the difference was attributed to the larger heat flux through the 50-mm specimens.

We ran another series of tests to determine if the hot and cold plates could be dynamically calibrated. The object of this series was reduced calibration time. These tests showed that the plate RTDs could not be dynamically calibrated to the desired accuracies due to the thermal mass of the plates.

At the conclusion of GHP work, we recalibrated the thermopiles in the bath. All four thermopiles repeated within acceptable accuracy and precision.

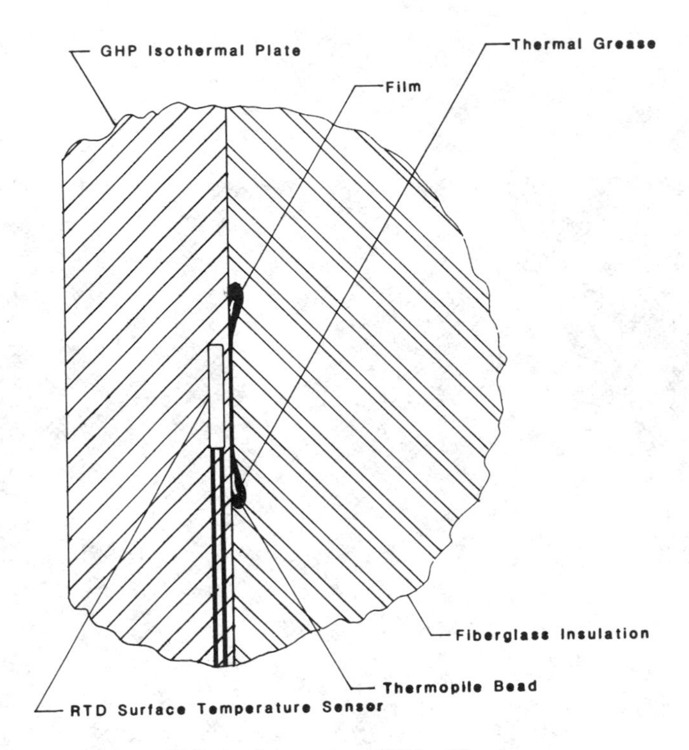

FIG. 4—*Schematic of GHP calibration.*

Acknowledgments

The author wishes to thank John Holman for his assistance in all phases of this work.

References

[1] *Manual on the Use of Thermocouples in Temperature Measurement, ASTM STP 470B*, American Society for Testing and Materials, Philadelphia, 1981.
[2] *SAS User's Guide*, Statistical Analysis System Institute, Raleigh, NC, 1979.

Summary

Summary

The late seventies and early eighties were a period of intense activity internationally in the general field of measurement of thermal performance of thermal insulating materials and systems. In North America, the 1976 revisions of the basic test method specifications had stimulated both new and improved measurement techniques and larger, more consistent and more sophisticated apparatus. In Europe, especially in Italy and France much work has taken place on improved designs of both types of apparatus. In France and more recently in Japan, considerable efforts were expended in the development of high sensitivity, faster, more responsive heat flux transducers. Finally in Belgium and Germany extensive work was on-going in improved computer modelling and analysis of the basic methods. Some work was being carried out also on both sides of the Atlantic, albeit at a much lower level, on the development and characterization of reference materials and transfer standards. After its formation in 1976, the ISO TC163 Committee on Thermal Insulation and in particular Subcommittee 1 on Test Methods adopted a similar approach to that taken in North America in 1976. After taking advantage of some of the later developments it was now becoming closer to the promulgation of several documents relating to measurements of thermal performance by standard hot plate and heat flow meter techniques without reference to a single apparatus design.

These activities prompted the decision by ASTM C16.30 Subcommittee to hold a meeting specifically addressing the state-of-the-art technology in these two most basic methods of measurement. Unlike the previous very successful C16.30 Symposia on heat transmission[1,2,3] and Thermal Insulation Conference[4] which addressed the respective subjects in more general context it was believed that the time was opportune for a more specialized meeting. This was particularly relevant to the activities of both the C16.30 Subcommittee and to the Working Groups 2 and 5 of Subcommittee 1 of ISO TC163. The former

[1]*Heat Transmission Measurements in Thermal Insulations, ASTM STP 544*, American Society for Testing and Materials, Philadelphia.

[2]*Thermal Transmission Measurements of Insulation, ASTM STP 660*, American Society for Testing and Materials, Philadelphia.

[3]*Thermal Insulation, Materials, and Systems for Energy Conservation in the '80s, ASTM STP 789*, American Society for Testing and Materials, Philadelphia.

[4]*Thermal Insulation Performance, ASTM STP 718*, American Society for Testing and Materials, Philadelphia.

was about to embark on a revision of the 1976 versions of ASTM C 177 and C 518 while the latter were in the final stages of developing documents on the international versions of the guarded hot plate and heat flow meter methods.

A meeting would provide the opportunity for the various international workers to discuss and share their expertise such that it would be most beneficial to all concerned both in reliable measurements and in Standards development. It would provide the opportunity to illustrate where disagreements exist and where future work is required.

In the end some seventy participants representing nine countries attended the two-day meeting. This indicated clearly the considerable national and international interest in the subject and the will to fulfill the goals of the meeting. Some twenty papers were presented, not all are included in this publication. Many lively and controversial topics were discussed in detail. As a result all who took part became much more aware of the problems and issues which affect reliable measurements by these methods. It is hoped that the participants will transfer this knowledge to their fellow workers. This should ensure that there will be general acceptance of results obtained by techniques which are based on sound principles documented in appropriate test specifications.

As in the earlier symposia and conferences the majority of the papers were devoted to analyses, use, and application of the methods at moderate temperatures. In most cases this was at or near room temperature and for materials used predominantly for buildings. This obviously reflects where the major emphasis of the work has been. However, it should not be forgotten that thermal insulations are utilized over a wide temperature range. Much of what was discussed at the meeting may well be totally valid at or near room temperature but could require extensive modification and refinement for general application under all conditions. Clearly this is an area where more work is required before there is complete understanding such that low error more precise measurements become the norm.

Existing test method documents relating to the two methods are based, in general, upon analyses undertaken some twenty to thirty years ago. Analytical techniques and tools have improved significantly since that time and more recently various workers have been applying them to the guarded hot plate method in particular. Much of this work can be applied to the heat flow meter method under certain defined conditions.

In order to take advantage of the efforts of these international workers, the first sessions addressed apparatus analysis and error analyses of the methods. It is only as a result of an understanding and verification of the sources of uncertain heat transfer paths within a basic apparatus that the experimentalist can establish the criteria whereby they may be eliminated or minimized.

Troussart began the discussion on this important subject with a contribution which summarized some of his efforts to analyze the sources of error in guarded hot plate measurements. His method of attack for the analysis was a three dimensional study of the plate using the finite element computational method. The author claimed a number of advantages over an exact mathematical solution. These included among others the fact that it addresses the subject without the need to resort to simplifying assumptions about gap temperature unbalance, materials, or boundary conditions. It can also account for nonuniform temperatures within various sections of the apparatus and for anisotropy. Finally for any given specimen the method provides a means to obtain the gap unbalance that corresponds to any guarding power input, thus assisting in design of a correct hot plate for a given performance or range of apparent thermal conductivities or both. The number, size, and type of temperature sensor wires across the gap between metering and guard sections can have a significant impact on the accuracy of measurement. In fact, his results indicate that precise measurements on a whole range of conductivities cannot be obtained with one apparatus utilizing one set of conditions.

In his paper Bode addressed the subject of the influence of the guard ring width on the accuracy of the measurements in the hot plate method. He investigated the temperature field in both square and circular plates and for both the double-sided and single-sided hot plate. His results show conclusively that the error is significantly dependent on the level of the surrounding or "ambient" temperature, and he defines the dependency such that it can be evaluated for an apparatus. Current test-method documents do address this subject, but the present results provide a better understanding of the problem. They provide more guidance for estimating the errors involved and the means for minimizing them for different plate types and sizes and test conditions. It is interesting to note that Troussart confirmed that his finite element model would produce the same results for the influence of the guard ring width.

The analysis by Eguchi deals with a one-sided circular guarded hot plate rather than the more general two-sided case. There is a distinct need for more attention to be paid to this case since for various reasons it is being utilized more often than heretofore. In many cases it is often not possible to obtain two "identical" pieces of an appropriate size of a material for a measurement. Thus, the experimenter is faced with applying a modified technique to evaluate one piece. Alternatively in undertaking a study on convection, for example, there is often a need to look at one specimen with heat flow in different directions. The one-sided method is often used for this purpose.

The present analysis investigated the total errors obtained with this type of plate by examining seven normalized parameters. These included the ratio of the radius of the guard ring to that of the metering plate, the thickness of the specimen, the thickness of the insulation between the hot plate and guard plate, and the edge insulation thickness. Other parameters included the ratio of the thermal conductivity of the specimen to that of the insulation between the other plates, the ratio of the direct heat flow across the gap to that from the specimen assuming unidirectional heat flow, and the ratio of the heat flow through the thermal bridges between the plates to that from the hot plate into the insulation again assuming one dimensional heat flow. Much experimental work is required to validate the findings. However, the analysis highlights the potential errors which may result in this measurement and provides the experimentalist with guidelines on how to minimize these errors.

In the last analytical paper, Rennex discussed and summarized the error analysis associated with the more recently developed 1-m diameter line-source heater form of guarded hot plate. This is a different concept of hot plate and was first proposed some twenty years ago by H. Robinson. As far as can be ascertained this type of hot plate has only been developed and used at National Bureau of Standards in the United States. It is to be hoped that other workers will consider building such apparatus since the technique can offer a number of potential advantages over the more conventional distributed heater version for homogeneous specimens especially in time of measurement.

The various measured contributions of parameters involved in an evaluation are examined for this plate. These include dimensional factors such as thickness of specimen metering area, temperature measurement and its control at positions in the plate, heat flows, edge losses, and plate emittance. There is no doubt that the large line source plate is capable of a very high accuracy of measurement over a limited temperature range up to 300 to 400 K providing it is designed and built very meticulously. It will be interesting to see if this technique is adopted by others for a wider temperature range than currently utilized. However, high cost of design and fabrication is a possible deterrent.

Discussion in these sessions was led by Hust of the National Bureau of Standards (NBS), Boulder, who contributed a brief presentation on precision and accuracy of the methods. It is clear that these are dependent on a large number of variables but especially temperature. Means are available to allow us to make some reliable estimates of the determinate errors but a careful series of measurements utilizing an apparatus on a variety of specimens is at different conditions needed to evaluate the indeterminate errors and thus the whole limits of its accuracy and precision.

This contribution stimulated much discussion on both practical considerations of the design of a hot plate and of an apparatus and how the idealized situations of the analyses can be attained or optimized. A particularly critical regime is the gap between the metering and guard sections of a plate. Performance is affected by the number, type, size, and distribution of thermocouple, heater wires across the gap, and the degree of unbalance across the gap. For specific plates at room temperature such as the NBS plate, it is possible to attain the order of 10 m K unbalance but for many practical plates, this is unattainable especially at elevated temperatures. Overall edge losses are also a factor. These are affected both by the degree of control of temperature uniformity of the plates and the external environment. Gradient guarding by coupling hot and cold plates was also discussed. There are general agreement that these areas need a great deal more research in order to quantify better the extent of edge effects.

A further discussion area involved the matter of the desirability of cutting the specimens such that the metering and guard areas are separated by a small gap filled with a packed loose-fill material to reduce convection and radiation in the gap. While this appears unnecessary in high resistance homogeneous thermal insulations, it is definitely recommended for layered composites, especially those with metal surfaces, hard, high conductivity materials, and for highly anisotropic specimens.

Some experimental evidence obtained from recent interlaboratory comparisons and limited Round-Robin programs in the United States seems to indicate that ±2 to 3% accuracy is possible with both methods over a very limited temperature range close to 300 K. More cooperative studies are ongoing under the auspices of ASTM C16, ISO TC 163, and the European Economic Community, respectively, both these are restricted to the room temperature range. The results of these studies are awaited anxiously to see if they corroborate the ±2% level.

Much more information is required for other temperatures, and for evacuated and other very high thermal resistance specimens. Separately, Pelanne and Tye reported some information on their findings at higher temperatures by each method. Once again, limited inter-laboratory comparisons have been involved. While it appears that two laboratories can attain agreement of ±5% or better with the hot plate at 1000 K results from other organizations extend this to ±12%. The heat flow meter method appears to be capable of attaining a ±5% or better accuracy at least to 500 K. However, there is an urgent need for more reference materials and transfer standards to be available both for calibration of heat meter apparatus and to assist in the verification of hot plate systems.

A major portion of the meeting was devoted, as could be expected, to the matter of apparatus design, operation, and use. A wide range of topics was discussed. These included various types of classical guarded hot plate and heat flow meter apparatus, the calibration and use of heat meter systems both in the laboratory and the field, and descriptions of two newer different types of hot plate apparatus each having potential advantages in simplicity of design and reduced cost.

In the first paper devoted to apparatus Brendeng described the construction, verification, and operation of both a large guarded hot plate and a heat flow meter apparatus for evaluating the overall thermal conductance of insulation panels for insulating the storage tanks for cryogenic liquids being carried in large ocean going carriers. Accurate boil-off characteristics for these tanks is a necessity in determining profitability of a carrier system. Thus, it is necessary to have suitable apparatus for evaluating accurately any proposed thermal insulation system and check boil-off of fluid. Norway, in particular, is a leader in this field, and the paper described the essential details of design and verification of their large 2 by 3 m apparatus which has been developed for evaluating different insulation assemblies, including evacuated systems, over the approximate temperature range 120 to 300 K.

In another paper from Europe, De Ponte discussed various ongoing research activities related to both methods. In particular he discussed the design and operation of the apparatus in relation to some of the guidelines provided by past and current analysis. He also included an outline of the current work of the Working Groups of ISO TC163 SC1 in the field.

Previous analytical work was based on solutions of individual error situations based on simplified configurations[5] are for "homogeneous" specimens. While these results are still valid the situations are more complex and involve interactions between the different sources. Luckily modern computational tools are now available to assist in the development of refined analysis. Practical matters involving the use of automatic control in order to reduce operation time and the effects of specimen material and size are also addressed. In particular the potential errors due to materials being anisotropic, being composite in form with high thermal conductivity surfaces or faces or of such a form that radiation is a predominant mode of heat transmission are discussed as are the mean to reduce the errors. This paper illustrated clearly various topics and issues which are the stimulus of further research and development.

For the case of the heat flow meter apparatus, the advantages and disadvantages of a single versus double heat flux transducer form are discussed. The author raised a rather important point in his presentation, in that this method is potentially capable of providing reproducibilities equal or better than that of the hot plate and accuracies closely the same as the hot plate, yet very little modelling and analysis has been undertaken specifically for the method. One is tempted to conclude that its relative simplicity is the reason for this phenomenon. However, his discussion indicates clearly that more attention should be paid to this area especially since the method is now so widely used worldwide both in the research laboratory and for quality assurance applications.

McElroy et al presented a paper describing in detail the design and operation of a large self-guarding heater plate apparatus. This work extends that which these workers started some years ago utilizing a special form of heater but for the cylindrical configuration and which was reported in an earlier Symposium.[2] The essential and somewhat unique feature of this method involves the use of a large very thin screen mesh heater of low heat capacity which is sandwiched between two specimens. An error analysis indicates that the measurement uncertainty over the range 300 to 330 K is less than 2% and experimental evidence shows that a repeatability and reproducibility of ±0.2% is possible. Further work is ongoing to investigate the use of this method for transient measurements to determine, under certain circumstances the apparent thermal diffusivity.

Another modified plate apparatus based on a design that has been in use for thin sheet materials for some 15 years was discussed by Hager. This

[5]*Symposium on Thermal Conductivity Measurements and Applications of Thermal Insulations, ASTM STP 217*, American Society for Testing and Materials, Philadelphia.

method again utilizes a very thin self-guarded rapid response heater. In this case, the heater is a metal foil which is so thin that lateral heat flow along the plane of the heater can be ignored. The author described the design and operation of this modified system for measuring thermal insulating materials and indicated that a 2% accuracy may be possible and 5% attainable. Once again, its potential advantages are ease of fabrication and operation and low costs.

In undertaking reliable measurements by the heat flow meter method one is utterly dependent upon stable, reproducible heat flux transducers, and reliable reference materials and transfer standards for calibration. In their paper, Bomberg and Solvason discuss these factors in detail. They discuss the many and varied sources of error and the particular drawbacks of the method. However, in comparing two methods of calibration, it is shown that by choosing to calibrate with adequate transfer standards it is possible to attain a high degree of accuracy. Conclusively, it is shown that one cannot rely solely on the use of an externally calibrated heat flux transducer in the apparatus. The authors conclude their informative paper by describing both the transfer standard and calibrated specimen bank established at NRC in Canada and a suggested test protocol which should be established to ensure that reliable measurements can be made.

The significant increase in need for measurements of thermal properties and performance of thermal insulations in the past two decades has stimulated development of commercial instruments. An array of reliable guarded hot plate and heat flow meter measurement systems covering various sample sizes, temperature ranges, environmental capabilities, and claiming conformance with appropriate national and international specification is now available to those not wishing to undertake the design and construction themselves. The paper by Coumou reviewed the currently available instruments from sources in the United States and Canada. He was able to include several from other countries but was not exhaustive in this respect due mostly to a lack of response from some commercial sources to his questionnaire. The present survey illustrated that while several different manufacturers had various models available one organization offers a flexible range of both hot plates and heat flow meter instruments to serve the overall needs of those in the insulation arena worldwide.

In addition to a purely factual account of the available equipment, the author discussed the often conflicting factors of what are the ideal technical requirements for an instrument and what the market will bear for cost. Overall, any instrument or system must be such that the operation, accuracy, and precision are consistent with current standards. It should be relatively simple and reliable in operation, versatile in design allowing for modification, attractive and functional, contain state-of-the-art components and naturally be cost effective. Some concluding remarks were made by this one particular member of a manufacturing organization concerning his view of future needs and developments in the area of commercial test and evaluation instrumentation. Clearly, the past decade has seen a distinct change in this particular area. Commercially designed and built equipment is now totally accepted within the insulation community both for routine quality assurance purpose and for the research and development laboratory.

Two papers by thermal insulation manufacturers followed each illustrating particular activities relating to the use of heat flow meter equipment including commercial equipment described earlier. In the first of these, McCaa described the critical parameters which are involved in testing of loose-fill thermal insulation. He illustrated how the heat flow meter can be used to examine the effects of these parameters and especially specimen preparation for low density materials. Careful apparatus calibration utilizing transfer standards of low density is essential for this purpose.

In the second industry paper, Pelanne described the development of an extensive heat flow meter apparatus calibration program established throughout his company. This author has worked extensively in this subject for many years, and his commendable efforts in trying to ensure precision and quality are well recognized. The program involved the careful selection of specimens of fiber glass of various densities, measurements on these at the National Bureau of Standards, followed by the establishment of a secondary set of reference specimens utilizing those evaluated at NBS. By continuous checking at each plant, coupled with interchange of specimens, the company can be assured to the best of its knowledge that all equipment is maintained for audit purposes at a requisite stage of calibration such that precise, reliable measurements are obtained at all plants, and truly valid comparisons are possible. This ensures that the products are maintained to the degree that satisfies requirements of Federal and State regulatory bodies.

The use of heat flux transducers has increased significantly in recent years, particularly in measurements of thermal performance of building materials and systems. While not an overwhelming market, their use in heat meter apparatus has increased. It is natural that larger, improved, more responsive heat flux transducers would be developed for specific purposes. In their use for heat meter equipment the most significant improvements were made by Klarsfeld and his colleagues in France and first reported in an earlier STP[2] of this series. Their transducer is based on a regular array of closely packed sensors as opposed to the strip or line form, and the surrounding guard area is of exactly the same material and form as the central metering area. Miyake and Eguchi in their paper describe recent development in Japan of another large area heat flux transducer. It is somewhat similar to the French type in that it is manufactured by means of photoetching and electroplating techniques. The authors described also a new calibration technique using both absolute and relative methods in one apparatus, thereby establishing easy cross checking and confirmation. Another advantage of this transducer is that it is flexible, and the large area can be cut into smaller areas and utilized as required particularly for field measurements. The availability of these various improved transducers is a necessary item to ensure that the heat meter method continues to be a valid and accurate method of measurement and that heat flux transducers can be used more reliably for other evaluation of materials and systems performance.

In the sessions on apparatus and instrumentation, three short presentations stimulated additional discussion topics. In the first, Klarsfeld described his recent work on the development of a double-guarded, single specimen hot plate apparatus for more rapid measurements at elevated temperatures. This is an extremely useful concept particularly for quality assurance purposes and is also being further investigated by Coumou and his colleagues. However, if transfer standards and high temperature heat flux transducers were available, conversion of the concept to a heat flow meter method would be a further beneficial development.

Sparrell described a two heat flux transducer version of heat flow meter apparatus. The use of two transducers may be beneficial to the overall measurement providing great care is taken to calibrate each separately at the operating conditions. Contrary to the widely accepted opinion that by using two heat flux transducers edge losses are eliminated Shirtliffe illustrated that this was not so. All that is accomplished is to reduce the effective thickness of the specimen to half its value. Other speakers related experiences of seeing very large differences between heat fluxes measured by the two transducers and that averaging of such amounts must provide large uncertainties in the final measured value. The German national standard places an 8% value as an upper acceptable level of difference in the measured heat flux.

Much discussion took place on the subject of equilibrium and measurement time particularly for the heat flow meter method. This is highly dependent on the properties and thickness of a specimen. It also depends on whether one is measuring similar specimens or a set of variable property and size specimens. In general, for the former case at 300 K, times vary between 10 and 150 min for specimens of 25 and 150 mm. For many purposes this is quite adequate but for quality control and assurance much short times are preferred. Means to accomplish shorter times were discussed especially the optimum position of control sensors. In addition, use of thin foil or mesh type heater methods may assist in this respect.

In the third short presentation, Degenne described improved forms of heat flux transducers developed originally in France. The current version is much thinner, has both a higher sensitivity and a much faster response time. Overall such developments coupled with those being undertaken in Japan are assisting significantly in improving our measurements by the heat flow meter method.

In the Methodology session, Schumann discussed the need for very accurate determinations of temperature in measurements by the guarded hot plate method. In this context, he described the method that was utilized to calibrate the individual sensors embedded in the plates of his large 1.2 m square hot plate. Because these sensors could not be removed from the plates for external calibration, a precise method of *in situ* calibration was devised. The requirement to be able to measure and know the surface temperatures to within 0.03 K was attained after three attempts with the precision and accuracy improving with each measurement. This paper illustrated a very important factor in overall measurement technique which is often neglected.

In the final paper, Larson described the use of a 1.2 m square heated temperature controlled plate containing heat flux transducers and its application to evaluate thermal performance properties of building components, particularly roofs in the field. Full details of the design calibration and operation of this unit were supplied. While not specifically a guarded hot plate or heat flow meter apparatus, the contribution illustrated that a thorough knowledge of the principles involved in these methods is necessary to extend measurements into the field.

Discussion in the methodology sessions centered upon two major areas, namely, reference materials and use of heat flux transducers particularly for systems and field measurements.

De Ponte made a short presentation on current activities within the EEC on their reference materials program. This includes a high density glass fiber board, Pyrex 7740 glass, a polystyrene cellular plastic, and a proposed one consisting of a low density blanket of hollow polymer fibers of uniform diameter. Powell added further to the subject by describing current activities at NBS in calibration of materials. The high density fibrous glass board is now a Standard Reference Material (SRM 1450b), a lower density blanket (1451) is being evaluated at cryogenic temperatures, and the line source plate is being used to measure individual transfer standards of fiber glass blankets of low densities. In addition, he mentioned that NBS in Boulder was in the process of designing and building a high temperature hot plate. This was for use up to 800 K in order to start investigating potential higher temperature reference materials. This latter need is becoming more critical since the use of insulations at elevated temperatures has increased significantly in recent years. Tye, in particular referred to the C16.30 efforts during the seventies to identify potential materials, and he and many others were disturbed at the lack of progress which has been made. This is probably the one major subject where the need for additional work is critical.

Discussion of heat flux transducer use centered upon the question of their reliability and accuracy. In addition, their adequate representation of heat flows within more complex heterogeneous systems and for field conditions was also questioned. This is a particularly salient point since, in general, small 51 mm diameter or square transducers are most widely used at the present time. Their simplicity of use is widely accepted, however, this is offset by a real uncertainty in the accuracy that is possible. The sessions and discussion periods highlighted a need for a future meeting to present a state of the art in heat flux transducer technology.

This particular subject is in its infancy and the accuracy and precision of measurements is well below that which can be demonstrated in the laboratory. Several issues and topics are controversial and need to be addressed by additional work. This last session provided the bridge between laboratory and field measurements. It is particularly significant that one year later a workshop on heat flux transducer technology was arranged by Subcommittee C16.30 to address this growing subject. The Special Technical Publication volumes containing the proceedings of both meetings should prove invaluable to all of those currently involved or planning involvement in thermal properties measurements on insulations and insulation systems.

It is now over 30 years since ASTM C16 on Thermal Insulation began the development of technical symposia and meetings as part of its objectives to ensure transfer of technology in the subject of thermal insulation performance. During that time, the subject matter has been both specific and general, and the whole area has grown more complex. However, as a result of these efforts, the subject of materials and systems performance is now better understood especially for materials in the building regime. These contributions, while significant in certain contexts, serve only to illustrate that much more work is required, especially at other temperatures, and for real life conditions and for real insulation systems.

We trust, therefore, that the reader is now stimulated to partake in this further work such that even more significant advances in measurement technology will result. Subcommittee C16.30 will continue in its efforts to ensure that the results and findings of such efforts are published in future volumes of this series and are included in new and revised test method specifications under its jurisdiction.

C. J. Shirtliffe

Division of Building Research, National Research Council, Canada K1A 0R6; editor.

R. P. Tye

Thermatest Department, Dynatech R/D Company, Cambridge, MA 02139; editor.

Index

U

W

Y

Z